Travel Disruptions

This timely and pivotal volume explores the nexus of a wide range of travel disruptions and their multifaceted implications on the global travel and tourism industry.

As the global travel and tourism industry struggles with an increasingly unpredictable landscape marked by natural disasters, health crises, geopolitical upheavals, and societal transformations, comprehensive insights and practical strategies are needed to help navigate these challenges. Comprising 13 specifically commissioned chapters authored by global experts in the field, this insightful and richly illustrated book therefore explores the interplay of unique factors influencing travel behaviour, policy responses, and the industry's overall resilience. With a multidisciplinary focus and international case studies throughout, the book examines innovative strategies for building resilience, post-disruption recovery mechanisms, and crisis communication protocols, as well as exploring the various social, cultural, and ethical implications often linked with economic growth and cultural preservation. It provides a global understanding of the ever-evolving landscape of travel disruptions, the psychological nuances that tend to guide traveller choices and a comprehensive assessment of policy responses, governance strategies, and stakeholder collaborations that aim to strengthen resilience in the industry, thus helping to ensure sustainable growth in the future.

This book is a pivotal resource for academics, researchers, industry professionals, and policymakers concerned with the ever-evolving nature of travel and tourism, going beyond theoretical discourse.

Emmet McLoughlin is a Senior Lecturer in Tourism and Events Management at Cardiff Metropolitan University. Dr McLoughlin is a member of the Welsh Centre for Tourism Research (WCTR) and a Fellow of the Higher Education Academy (UK). His research interests include destination and event management, leisure studies, and sociology. His current research is focused on destination resilience, leisure and biosecurity, and crisis management in the events sector.

Domhnall Melly is an Assistant Lecturer in Tourism in the Department of Marketing, Tourism, and Sport at ATU Sligo, Ireland. Domhnall engages in research across a wide range of current issues in the broader tourism industry, particularly concerning tourism and biosecurity, tourism sustainability, tourism policy and planning, and festival and event management.

James Hanrahan is a Lecturer in Tourism Management and a Member of the Sustainable Tourism Observatory (STORY) group at ATU Sligo, Ireland. His lines of research are mainly focused on planning for sustainable destination management as well as community participation in the tourism planning process. Current industry-funded research is focused on the drive tourism route "The Wild Atlantic Way" and the tourist's use and adoption of Travel Apps and mobile devices.

Contemporary Geographies of Leisure, Tourism and Mobility
Series Editor:
C. Michael Hall
Professor at the Department of Management, College of Business and Economics, University of Canterbury, Christchurch, New Zealand

The aim of this series is to explore and communicate the intersections and relationships between leisure, tourism and human mobility within the social sciences.

It will incorporate both traditional and new perspectives on leisure and tourism from contemporary geography, e.g. notions of identity, representation and culture, while also providing for perspectives from cognate areas such as anthropology, cultural studies, gastronomy and food studies, marketing, policy studies and political economy, regional and urban planning, and sociology, within the development of an integrated field of leisure and tourism studies.

Also, increasingly, tourism and leisure are regarded as steps in a continuum of human mobility. Inclusion of mobility in the series offers the prospect to examine the relationship between tourism and migration, the sojourner, educational travel, and second home and retirement travel phenomena.

The series comprises two strands:

Contemporary Geographies of Leisure, Tourism and Mobility aims to address the needs of students and academics, and the titles will be published simultaneously in hardback and paperback.

Routledge Studies in Contemporary Geographies of Leisure, Tourism and Mobility is a forum for innovative new research intended for research students and academics, and the titles will initially be available in hardback only.

Titles include:

Cultural Heritage on the Urban Peripheries
Towards New Research Paradigms
Edited by María García Hernández and Maria Gravari-Barbas

Travel Disruptions
Impacts, Responses and Resilience
Edited by Emmet McLoughlin, Domhnall Melly and James Hanrahan

For more information about this series, please visit: www.routledge.com/Contemporary-Geographies-of-Leisure-Tourism-and-Mobility/book-series/SE0522

Travel Disruptions

Impacts, Responses and Resilience

Edited by Emmet McLoughlin, Domhnall Melly and James Hanrahan

LONDON AND NEW YORK

First published 2025
by Routledge
4 Park Square, Milton Park, Abingdon, Oxon OX14 4RN

and by Routledge
605 Third Avenue, New York, NY 10158

Routledge is an imprint of the Taylor & Francis Group, an informa business

British Library Cataloguing-in-Publication Data
A catalogue record for this book is available from the British Library

ISBN: 978-1-032-72053-1 (hbk)
ISBN: 978-1-032-72054-8 (pbk)
ISBN: 978-1-032-72055-5 (ebk)

DOI: 10.4324/9781032720555

Typeset in Times New Roman
by SPi Technologies India Pvt Ltd (Straive)

Contents

Figures

Tables

Contributors

Fatemeh Bagheri completed her bachelor's degree in Tourism Management, master's degree in Tourism Management specializing in Tourism Marketing, and PhD in Tourism Management. She is a Postdoctoral Researcher at the University of Algarve, where she is involved in various research projects. Fatemeh is also an Integrated Member of the Research Center for Tourism, Sustainability and Wellbeing (CinTurs). She has authored papers and served as an editorial board and reviewer for several top-tier journals specializing in tourism and hospitality.

Bee Lian Song is a Senior Lecturer at the Asia Pacific University of Technology and Innovation, Malaysia. She has 10 years of industrial experience and more than 12 years of experience in academia. Her research interests include digital marketing, marketing sustainability, consumer behaviour, and corporate social responsibility. She has published various articles in reputable Web of Science, SCOPUS, and ABDC-indexed journals.

Danijela Boljat holds a master's degree in Tourism and Hospitality from the Faculty of Economics in Split. She is currently employed at an online travel agency with strong ties to the local IT and marketing sectors. Her professional interests focus on regenerative tourism, managing overtourism, and the impact of digitalization and technology on the tourism industry. With a respective academic foundation and practical experience in tourism development, she brings a comprehensive and forward-thinking perspective to the evolving challenges and opportunities in the field.

Liping A. Cai is a Professor in the White Lodging-J.W. Marriott, Jr. School of Hospitality and Tourism Management at Purdue University.

Neil Carr's research focuses on power, freedom, and wellbeing within tourism and leisure experiences, with a particular emphasis on children and families, sex, and animals (especially dogs).

Jonathon Day is an Associate Professor in the White Lodging-J.W. Marriott, Jr. School of Hospitality and Tourism Management at Purdue University.

Elyssa Fanning graduated from the University of Sussex with a bachelor's degree in law with Politics. She is an Account Executive at Silverstone Communications and assists with political research for client projects.

Shameena Fernando's research focuses on disaster management and tourism, specifically the ripple effect of a three-decade civil war, tsunami, Easter-attacks, and COVID-19 on small and medium accommodation operators in Sri Lanka.

Siao Fui Wong currently works at Modul School of Tourism and Hospitality Management Nanjing, Nanjing Tech University Pujiang Institute, China. Siao Fui does qualitative research and focuses on online destination image, destination branding, and sense of place.

Kathryn Hayat is an Associate Professor and Head of the Department of Hospitality and Tourism at University College Birmingham, England. Her experience includes over a decade of lecturing tourism and aviation management, specialising in digital marketing and social media. Before her career in academia, Kathryn worked in digital marketing and social media roles at tour operators and airline wholesalers, pioneering the use of Facebook for both B2C and B2B purposes.

Andrew Higgins is a PhD candidate at Atlantic Technical University Sligo, where he intends to publish his research in 2025 titled "Development of an Urban/Rural collaborative framework as a tool for regional destination branding." Andrew has been researching collaborative networks and destination branding since his undergraduate studies. He has conducted industry research for stakeholder networks and DMOs and has published conference papers at Irish conferences such as the Tourism Hospitality Research in Ireland Conference (THRIC) and the Irish Academy of Management Conference (IAM) as well as international conferences such as the International Conference on Tourism and Business (ICTB).

Adam Jones, PhD, is a Principal Lecturer of Strategy and Marketing at the University of Brighton. He started his career in the travel industry working as a tour guide rising to the position as Head of Marketing. It was during his time as Head of Marketing, Research and Product Development that he developed an expert understanding of the issues around strategic direction and the challenges of applying the principles of strategic marketing, PR and crisis communication in an extremely competitive and challenging commercial environment. His research interests centre around sustainability and inclusivity especially applied to Tourism, Events and Hospitality.

Christin Khardani is a PhD student and Lecturer at the Economic Geography Group at the Ludwig-Maximilians-Universität München, Germany. She also holds a scholarship from LaKoF (State Conference of Women's and Equal Opportunities Officers at Bavarian Universities). Her research focus is on crisis and disaster management and safety and security in tourism, specifically in the industry of tour operators and in tourism destinations. Within this scope, she also focuses on multi-stakeholder collaboration.

Xinran Lehto is a Professor in the White Lodging-J.W. Marriott, Jr. School of Hospitality and Tourism Management at Purdue University.

Wong Ling Chai is the Digital Marketing Lecturer at the Asia Pacific University of Technology and Innovation (APU). His responsibilities involve teaching retail marketing management, global marketing, consumer behaviour, integrated marketing communication, cyber marketing, and e-commerce. He supervises PhD candidates, actively guides MBA students in dissertation writing, and serves as an internal examiner. Dr. Wong is also an external examiner at North Borneo University College. He has been actively involved in research, specializing in mobile marketing, consumer behaviour, and sustainable marketing practices, recognized as an Outstanding Reviewer in 2018.

Sarasadat Makian is a junior researcher focusing on tourism studies. She received her Doctorate in Geography from the University of Grenoble Alpes (UGA) in France. Her academic career started in Iran, where she completed her Tourism Management bachelor's and master's degrees. She pursued additional specialization and, at UGA, earned a second master's degree in Tourism, Innovation and Transition, demonstrating her constant commitment to expanding the body of knowledge in her field. Her research primarily focuses on future tourism studies and tourism destinations' future-based development. In addition to her publications, she also serves as a reviewer for several journals.

Ante Mandić, PhD, is an Assistant Professor of Sustainable Tourism at the University of Split's Faculty of Economics, Business, and Tourism in Croatia. He is affiliated with Colorado State University and serves as manager of SmartCulTour Living Lab and Associate Editor of the Journal of Ecotourism. Professor Mandić's research focuses on sustainable tourism in nature-based destinations. He has published extensively and has worked on international projects funded by INTERREG MED and HORIZON 2020.

Vanda Maráková is a Full Professor at the Faculty of Economics, Matej Bel University, Slovakia, with over 20 years of experience in tourism studies. Her research focuses on tourism management, marketing communication, corporate social responsibility, sustainability, destination competitiveness, and innovation. She has contributed to 40 national and international research projects, resulting in numerous published works. Nominated by the Slovak Business Agency, she serves on the High-Level Advisory Board EU ECO-TANDEM in the area of finance and sustainable tourism development. She is also an active member of the Triangle Knowledge Alliance, dedicated to sustainable tourism in Europe's protected areas, and a member of AIEST.

Sbusiso Mbuyane is a Tourism Management master's graduate from Tshwane University of Technology, Pretoria. His master's dissertation focused on travel agencies and their crisis management strategies. He continues to contribute to his academic profile through academic articles and book chapters. He worked as a travel practitioner from 2019 to 2021, before pursuing a career in academia. Sbusiso has a strong interest in studies of Tourism Management, Travel Agency Sector and digital technology within the travel and tourism industry.

Catherine McGuinn is a Lecturer of Marketing at Atlantic Technological University (Sligo Campus), a graduate member of the Marketing Institute of Ireland, and a member of the Irish Academy of Management. Catherine is the Director of the Centre for Research in the Social Professions (CRISP). Her research includes include branding, customer service, and the tourism sector.

Conor McTiernan is a Lecturer in the Department of Tourism and Sport at Atlantic Technological University, Donegal. His research interests include sustainable tourism policy formation, pro-environmental knowledge management, and the role of trust in inter-organisational and inter-personal tourism networks. Conor is a work package leader on REMODEL, a Horizon Europe capacity-building project with Bursa Uludag University and Universidad de Leon, and contributes to EU Green, a transnational European alliance of universities committed to embedding sustainability in peripheral economies. He regularly reviews for journals.

Lim Ming Fook is a Senior Lecturer at the Faculty of Business, Economics, and Accountancy, Universiti Malaysia Sabah, Malaysia. He has over eight years of experience in academia, focusing on teaching Marketing and Research Methods. His research interests are in digital marketing and consumer behaviour.

Rosa Naudé-Potgieter completed her PhD in Tourism Management at the North-West University in Potchefstroom, South Africa. Her thesis focused on the Quality-of-Work-Life of casino employees in South Africa. She completed her MBA at the Henley Business School. Rosa has been working in the casino industry since 2010, with most of her experience in the Operations, Marketing and Business Intelligence departments. In 2021, Rosa started lecturing at the Tshwane University of Technology, Pretoria, in the Tourism Department. Rosa has a keen interest in studies of Tourism Management, Casino Management, Labour Relations, Quality-of-Work-Life, eSports events, and Human Resource Management.

Annmarie Nicely is an Associate Professor in the White Lodging-J.W. Marriott, Jr. School of Hospitality and Tourism Management at Purdue University.

Ciarán Ó hAnnracháin has experience in Hospitality and Tourism Management education in the Irish, UK, US, and Swiss education systems. He has held academic management roles in Switzerland and Ireland and contributed to national advisory panels in both Hospitality and Tourism Higher Education. Ciarán was the Project Manager for Academic Planning and Strategy during the recent creation of Atlantic Technological University. His research interests include Higher Education Policy, Internationalisation of HE, Talent Management, and Staff Training and Development. He is Chair of the Operational Board of the World Technology University Network.

Aida Pinos-Navarrete is a Lecturer in the Department of Human Geography at the University of Granada (Spain). Her research focuses on health tourism (thermalism), well-being, and local development. She has published numerous papers and book chapters on spa tourism. In addition, she has participated in international research projects funded by European organizations and has been a visiting researcher at RWTH Aachen University (Germany) and the University of Exeter (United Kingdom). She has worked for institutions such as UGR Solidaria and the National Geographic Institute (Madrid).

Tee Poh Kiong is a Senior Lecturer in the School of Marketing and Management at Asia Pacific University in Kuala Lumpur, Malaysia. He has more than ten years of experience in both industry and academia. He has been actively involved in research and publication. His research interests include employability and career management, consumer behaviour, digital transformation, and sustainability marketing.

Aylin Poroy Arsoy holds a bachelor's degree in Business Administration, a master's degree in Accounting and Finance, and a doctorate in Accounting and Finance from Bursa Uludag University. Since joining BUU FEAS, she has progressed through the academic ranks, becoming a Professor in 2016. Aylin has extensive experience in cross-cultural studies and has led various national and international projects. Her academic research focuses on financial reporting, corporate transparency, corporate governance, accounting education, and auditing. She lectures at undergraduate, master's, and PhD levels, specialising in accounting and finance is actively involved in EU projects.

Daniela Garbin Praničević, PhD, is a Full Professor of Business Informatics at the Faculty of Economics, Business and Tourism, University of Split, Croatia. She teaches courses in ICT Project Management, Change Management, and Smart Tourism and Hospitality. Her research interests include the digital economy, sustainable management, and digital transformation in tourism. Dr. Praničević has contributed to several research projects and has published numerous papers based on project outcomes, reflecting her expertise in integrating digital transformation with sustainable practices in the tourism and hospitality sectors.

Ciara Quinlan is a Lecturer in the Department of Tourism and Sport at Atlantic Technological University, Donegal. She has been involved in Tourism and Tourism education for over 30 years. Her interests include sustainable destination development, marketing, and service experience. Ciara has developed and delivered tour guiding programmes. She maintains close links with industry stakeholders through her membership in the national tour guide working group.

Genka Rafailova is an Associate Professor of Tourism, Leisure, and Event Marketing and Management at the University of Economics, Varna, Bulgaria, where she also leads the Centre for Cultural, Creative, and Social Entrepreneurship. She holds a PhD in Tourism with a focus on sustainable development and has served as an expert on several EU-funded international projects, as well as EEA-funded initiatives. Her research interests include sustainable and smart tourism development, creative destinations, the creative industries, and the blue economy. Her latest publication is *Development of a Smart Tourist Destination 2030*. She is also a member of the Varna Tourism Chamber.

Laleh Ramezani is a Conservation Architect, Adaptive Reuse Designer, educator, and researcher. Holding a PhD and MA from the University of Tehran, her expertise spans the interdisciplinary domains of conservation, architecture, adaptive reuse of the built environment, and their impact on tourism. During her PhD, Laleh conducted research on the adaptive reuse of Iran's built environment in alignment with tourism development. She has been involved in significant projects such as the adaptive reuse of historical buildings in old Tehran, Gorgan, and Shiraz. Laleh also played a key role in the inscription of the Hawraman cultural landscape as a UNESCO World Heritage site.

Geri Silverstone, Founder and CEO of Silverstone Communications, is a political communications and community engagement strategy expert with 20 years of global experience across the built environment, transport infrastructure, education, and local government sectors. Geri's strengths lie in developing and implementing high-level fully integrated communications, public affairs, and stakeholder engagement strategies.

Shweta Singh graduated from the White Lodging-J.W. Marriott, Jr. School of Hospitality and Tourism Management at Purdue University.

Portia Pearl Siyanda Sifolo is a seasoned tourism expert with strong academic and practical experience. She has a Doctorate in Business Administration, focusing on supply chain management and stakeholder engagement in tourism. Her research interests include digital collaboration in African tourism and youth entrepreneurship. She has published extensively and holds various leadership positions in the travel and tourism industry.

Ying Ying Tiong serves as an executive committee member of the Social Innovation Movement Association and is the Associate Dean, Research and Development, and Senior Lecturer at the Faculty of Business, Curtin University Malaysia. Her research interests are centred around green marketing strategy, entrepreneurship, consumer behaviour, and responsible tourism. Her notable achievements include publications, successful acquisition of research grants, active involvement in government projects, and engagement with various industry stakeholders.

Michał Żemła is an Associate Professor of Tourism Management at the Department of Management in Tourism and Sport, at the Institute of Entrepreneurship at Jagiellonian University in Krakow, Poland. His recent research interests include overtourism in European cities, the experience economy in tourism, the development of winter sports destinations, and the sharing economy. He published several papers and chapters in esteemed journals and books.

Introduction

Exploring Travel Disruptions: Impacts, Responses, and Resilience

Emmet McLoughlin, Domhnall Melly and James Hanrahan

With increasing disruptions from natural disasters, health crises, and geopolitical events, understanding the multifaceted impacts on the travel ecosystem has become essential. Analysing the economic, sociocultural, and environmental consequences across interconnected destinations is critical to identifying systemic risks and vulnerabilities (Nyaupane, Poudel, & Timothy, 2018; Alvarez, Bahja, & Fyall, 2022). Equally crucial is the exploration of response mechanisms, including crisis management, coordinated communication, and government policies designed to maintain traveller confidence and facilitate recovery (Carlsen & Liburd, 2008). However, as Calgaro, Lloyd, and Dominey-Howes (2014) argue, simply mitigating immediate impacts is insufficient; a proactive, holistic approach is required to build long-term resilience within travel systems.

Resilience necessitates viewing travel as an integrated human-environment system, with an emphasis on leveraging flexibility, collaboration, and innovation to foster transformative adaptation (Cheer & Lew, 2017; Bui et al., 2020). This involves not only contingency planning and crisis training but also the promotion of stakeholder dialogue and coordination to effectively manage risks. By integrating insights on impacts, responses, and resilience, stakeholders can develop actionable policies that enhance the capacity to navigate crises while safeguarding cultural heritage, economic stability, and environmental integrity. In an increasingly interconnected global landscape, building collective resilience is indispensable for ensuring sustainable and thriving tourism into the future. The contributions within this edited collection, "Exploring Travel Disruptions: Impacts, Responses, and Resilience", aim to offer further clarity on the complex landscape of travel disruptions, providing critical insights for researchers, practitioners, and policymakers alike.

The role of heritage conservation and shared social identity in the context of tourism resilience is explored in the first chapter where Fatemeh Bagheri, Sarasadat Makian, and Laleh Ramezani investigate the connection between heritage conservation and tourism development, focusing on how preserving cultural heritage fosters a shared social identity that strengthens community resilience. The authors utilised three Iranian case studies to demonstrate how successful conservation efforts can bolster local tourism by building a stronger community identity and resilience against disruptions.

DOI: 10.4324/9781032720555-1

The issue of global tourism in the wake of COVID-19 is examined in the chapter by Danijela Boljat, Daniela Garbin Praničević, and Ante Mandić, who offer a bibliometric analysis of global research on the impacts of COVID-19 on tourism, highlighting key areas such as crisis management, policy responses, and changes in traveller behaviour. Their analysis identified emerging trends in pandemic-related tourism research together with examining collaborative efforts across countries and institutions.

Regarding responses to travel disruptions, the third chapter by Christin Khardani focuses on stakeholder collaboration in destination crisis management. In particular, this chapter examines the role of key tourism stakeholders, particularly tour operators and Destination Management Organisations (DMOs), in managing tourism crises. More specifically, this chapter highlights the gap between theoretical models and real-world practices, with the author proposing strategic collaboration among stakeholders as the most appropriate way to enhance destination resilience in times of crisis.

The chapter by Conor McTiernan, Ciara Quinlan, Ciarán Ó hAnnracháin, and Aylin Poroy Arsoy explores how travel disruptions influence tourists' trust in tour guides with the authors emphasising the importance of trustworthiness as a factor in ensuring business continuity during crises. Moreover, this chapter then delves into the risks that tour guides face and suggests ways to foster trust in the face of travel disruptions.

The evolving use of AI-driven automation in travel decision intelligence through periods of disruption is then investigated by Kathryn Hayat who discusses the increasing use of AI and automation in the travel industry, particularly during periods of disruption. This chapter further explores how AI tools are being used to improve decision-making processes for travellers and travel agencies, thus helping to anticipate and mitigate the impacts of crises on tourism.

The sixth chapter by Adam Jones, Elyssa Fanning, and Geri Silverstone focuses on communication strategies during aviation crises, such as technical failures or airport closures. Here the authors analyse and discuss how airlines can effectively manage public relations and customer expectations through transparent and timely communication in times of crisis, helping to maintain brand reputation and customer trust.

Navigating post-pandemic destination marketing communication is the focus of the seventh chapter by Shweta Singha, Annmarie Nicely, Jonathon Day, Liping A. Cai, and Xinran Lehto. Drawing from both crisis communication and disaster management theories, this chapter examines how destination marketing organizations (DMOs) can effectively communicate with tourists post-pandemic. Furthermore, the authors provide a framework for integrating crisis communication into destination marketing strategies to rebuild trust and stimulate tourism recovery.

The connection between travel disruptions and the challenges faced by travel agencies during crises is explored by Sbusiso Mbuyane, Portia Pearl

Siyanda Sifolo, and Rosa-Anne Naude-Potgieter in the eighth chapter. By focusing on their responses and strategies for building resilience, the authors here discuss the role of digital transformation and stakeholder collaboration in mitigating the negative effects of travel disruptions, with examples from global practices and a special focus on South Africa.

In discussing the various issues around building resilience in travel, Vanda Maráková, Michał Żemła, and Genka Rafailova provide a theoretical overview of resilience frameworks in the travel and tourism industry. More specifically, the authors outline several key concepts related to resilience such as adaptability and recovery and discuss how these frameworks can be applied to enhance the tourism sector's ability to withstand and bounce back from crises.

Chapter 10 by Shameena Fernando and Neil Carr examines the concept of multistakeholder resilience building. More specifically, the authors explore the role of various stakeholders, including governments, businesses, and local communities, in building resilience within the tourism sector. This chapter emphasises the importance of collaboration and outlines strategies for aligning the interests of diverse stakeholders to create a more resilient tourism system.

Focusing on the operational aspects of resilience, authors Ling Chai Wong, Siao Fui Wong, Bee Lian Song, Ying Ying Tiong, Tee Poh Kiong, and Ming Fook Lim define the capabilities required by travel systems to effectively respond to disruptions. This chapter identifies key resilience-building factors, such as flexibility, adaptability, and technological integration, that can help travel organisations navigate crises more effectively.

How DMOs responded during COVID-19 lockdowns and the implications on stakeholder engagement are the focus of Chapter 12 by Andrew Higgins and Catherine McGuinn. Here the authors examine the responses of DMOs during the COVID-19 lockdowns, focusing on how they adapted their strategies to engage stakeholders and support the tourism industry's recovery. This chapter highlights the lessons learnt from the pandemic and suggests future strategies for improving stakeholder engagement during crises.

The concept of resilience within the context of Spanish spa tourism is examined by Aida Pinos-Navarrete, who discusses how this niche sector has responded to crises by adapting its services and experiences. Furthermore, this chapter offers proposals for enhancing the resilience of spa tourism through innovative strategies and improved stakeholder engagement.

The guest editors hope this collection illustrates the connections between travel disruptions, their impacts, responses, and resilience. Through case studies and analyses, the chapters examine key response mechanisms, stakeholder collaboration, and resilience-building strategies, aiming to foster long-term sustainability in the travel sector. The editors further hope these contributions shed light on how travel stakeholders can navigate crises, protect cultural and environmental assets, and strengthen resilience in an interconnected world.

References

Alvarez, S., Bahja, F., & Fyall, A. (2022). A framework to identify destination vulnerability to hazards. *Tourism Management, 90*, 104469.

Bui, H. T., Jones, T. E., Weaver, D. B., & Le, A. (2020). The adaptive resilience of living cultural heritage in a tourism destination. *Journal of Sustainable Tourism, 28*(7), 1022–1040.

Calgaro, E., Lloyd, K., & Dominey-Howes, D. (2014). From vulnerability to transformation: A framework for assessing the vulnerability and resilience of tourism destinations. *Journal of Sustainable Tourism, 22*(3), 341–360.

Carlsen, J. C., & Liburd, J. J. (2008). Developing a research agenda for tourism crisis management, market recovery and communications. *Journal of Travel & Tourism Marketing, 23*(2–4), 265–276.

Cheer, J. M., & Lew, A. A. (2017). Understanding tourism resilience: Adapting to social, political, and economic change. In Joseph M. Cheer, & Alan A. Lew (Eds.), *Tourism, resilience and sustainability: Adapting to social, political and economic change* (pp. 3–17). Routledge.

Nyaupane, G. P., Poudel, S., & Timothy, D. J. (2018). Assessing the sustainability of tourism systems: A social–ecological approach. *Tourism Review International, 22*(1), 49–66.

1 Exploring the Role of Heritage Conservation and Shared Social Identity in Tourism Resilience

Fatemeh Bagheri, Sarasadat Makian and Laleh Ramezani

1.1 Introduction

Heritage holds significant importance for enhancing communities' social and cultural well-being (Otero, 2022; Tweed & Sutherland, 2007). Following the World Heritage Convention established by UNESCO in 1972, the term "cultural heritage" encompasses a collection of structures or locations with historical, aesthetic, archaeological, scientific, ethnological, or anthropological significance (Sotirova et al., 2012). Cultural heritage is considered an invaluable socio-economic resource (Baglioni et al., 2021). Therefore, the primary objective of heritage conservation extends beyond safeguarding history and culture for the future (Otero, 2022) and involves establishing and upkeeping shared social institutions within societies (Mydland & Grahn, 2012). Indeed, heritage plays a crucial role in terms of social value (see Parga-Dans et al., 2020), a sense of belonging to a place, and ultimately contributes to the shared social identity of communities (e.g., Escalera-Reyes, 2020). The significance of this matter becomes particularly pronounced in traditional societies, which are commonly understood to derive their fundamental characteristics, values, and identities from their historical roots (Kideghesho, 2008). Therefore, when discussing shared social identity, tourism development resulting from heritage conservation carries substantial importance.

On the other hand, building and strengthening connections between people and places can contribute to community and tourism destination resilience. Tourism destination communities should enhance their resilience by cultivating diverse forms of community capital (Wakil et al., 2021), including social identity. Consequently, heritage tourism development can enhance the social identity within societies (Ruiz Ballesteros & Hernández Ramírez, 2007), thus strengthening the tourism sector's resilience.

In tourism, resilience can be understood as the capacity to utilise community resources (ranging from cultural heritage to social identity) to flourish in an environment characterised by change, risk, and uncertainty (Bec et al., 2016). Residents' attitudes towards the tourism phenomenon are substantially influenced by their social identity, thereby influencing their support for tourism development (Nunkoo & Gursoy, 2012) and the destination's resilience. This is

DOI: 10.4324/9781032720555-2

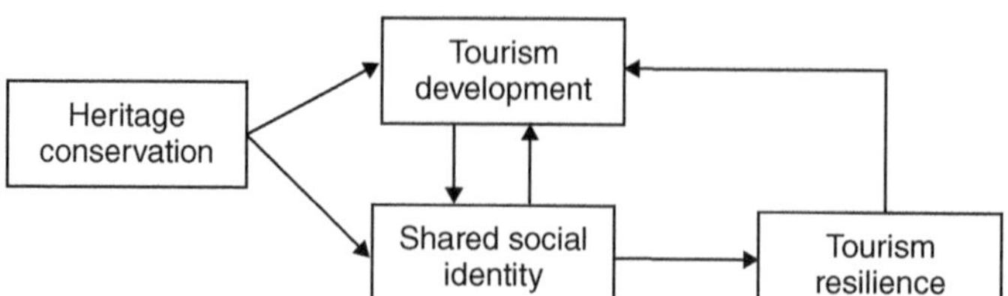

Figure 1.1 Conceptual framework of the study.

essential to fortify the mitigation of disruptions' effects and ensure that the tourism industry recovers sustainably at the community level.

Focusing on social identity and resilience, this study conceptually explores the relationship between heritage conservation and tourism in increasing the sense of shared identity and examines its implications for tourism resilience. To our knowledge, this is the first study to investigate the relationship between heritage conservation, tourism development, social identity, and tourism resilience concepts. Therefore, it has the potential to foster advocacy for tourism resilience by emphasising the importance of heritage conservation and the community's social identity.

This chapter will analyse three different case studies from Iran, as a traditional society, to further illustrate the theoretical concepts we are examining (Figure 1.1). This study offers valuable insights into the collaborative potential of heritage conservation and tourism, demonstrating how they can synergistically foster social identity and enhance the tourism sector's resilience. These findings provide applicable lessons not only in Iran but also in other communities worldwide.

1.2 Literature Review

1.2.1 Heritage Conservation, Tourism, and Shared Social Identity

Heritage is recognised as one of the most universal resources for tourism (Di Pietro et al., 2018). Cultural heritage sites, such as historical buildings or cultural landmarks, are widely acknowledged for their remarkable universal values that contribute to advancing tourism activities in their respective locations. Additionally, these sites play a key role in social life by constructing notions of individuality and group identity (Dela Santa & Tiatco, 2019). Tourism has the potential to provide a financial basis for the preservation and management of heritage sites. On the other hand, conservation initiatives can contribute to preserving the authenticity of heritage tourism seeks to capitalise on (Chapagain, 2016). Conservation is the process of looking after, interpreting, and managing heritage to ensure its safekeeping for future generations and its contribution to the longevity of a nation's heritage and symbolic identity (Bagheri et al., 2022). However, there are risks related to authenticity loss and impoverishment of cultural identity or a sense of belonging felt by local inhabitants that emerge with over-tourism in cultural heritage.

Cultural heritage can also stimulate local socio-economic development and foster a sense of local identity, pride (Cerisola & Panzera, 2024), belonging, and citizenship value. Moreover, the utilisation and conservation of cultural heritage can foster solidarity, social inclusivity, and innovation, thereby aiding in preserving historical legacies and transmitting traditions to subsequent generations. It heightens locals' consciousness and augments their ability to identify with heritage-related circumstances (Di Pietro et al., 2018). Meanwhile, as highlighted by Makian et al. (2022), to facilitate tourism development, it is imperative to actively engage local community members and promote the creation of community identity. This enhances the community's overall well-being.

In this regard, shared social identity refers to a collection of individuals who perceive one another as part of a shared social group, fostering a sense of "we-ness" (Neville et al., 2022) through the development of shared feelings of belonging and attachment (Escalera-Reyes, 2020). Developing a shared social identity results in increased trust, respect, collaboration, assistance, and support among group members, ultimately fostering a sense of intimacy. Shared identity is the foundation of successful groups and social actions (Neville et al., 2022). Hence, when examining heritage tourism, it is imperative to consider its role in social identity improvement as a fundamental feature. Identities possess symbolic, open, political, and dynamic community characteristics (Ruiz Ballesteros & Hernández Ramírez, 2007).

1.2.2 Shared Social Identity and Tourism Resilience

Based on the social identity framework, the identity may encompass various aspects of a place (Mansour et al., 2023); the place could be a city, a village, or a cultural heritage site. A strong collective identity linked to living spaces generates a profound sense of belonging, solidifying the interconnections between identity, space, and community. In this regard, communities that incorporate their identity into their destinations tend to receive greater support from locals, which can lead to increased levels of sustainability (Jiménez-Medina et al., 2021). Additionally, shared social identity catalyses the resilience process by shaping collective cognition and action. This identity increases the well-being of the group and its individuals and is linked to mechanisms that maintain group cohesion (Erfurth et al., 2021). By fostering cognitive, relational, and emotional changes, shared social identity enables individuals to effectively collaborate, constituting a source of social power and motivating collective action and resilience among residents during emergencies and beyond (Erfurth et al., 2021).

Resilience is characterised by the ability to adapt to and endure significant challenges, resulting in the development of new planning approaches that foster sustainable living for residents of affected areas (Jiménez-Medina et al., 2021). Therefore, tourism resilience aims to preserve the standards of tourism activities and overall quality of life in a tourism community at the desired level. Resilience is not an objective in itself but rather depends on the perceptions and actions of community members. On the other hand, enhancing the resilience of the tourism industry is vital for promoting sustainable tourism development (Guo et al., 2018). It means that tourism resilience refers to communities' ability to not only handle the negative impacts of tourism but also actively work towards achieving sustainable goals (Bui et al., 2020).

According to Wakil et al. (2021, p. 4), "an affective unit of identity and belonging" is one of the main concepts for evaluating communities. Shared social identity highlights the crucial role of collective identity in promoting a sense of belonging and supporting the communities' members within tourism destinations. The resilience of tourism development can increase in the face of both ordinary and unexpected shocks, as it is closely intertwined with the social and collective identity of communities and influenced by the preservation of cultural heritage. Enhancing the resilience of tourism is of utmost importance for guaranteeing long-term sustainability and safeguarding the inherent attributes of tourism destinations. As shown in Figure 1.1, these concepts are interrelated, highlighting that community members' roles are indispensable in shaping the trajectory of resilience and sustainable tourism.

To illustrate the real-world interconnections between the concepts of heritage conservation, tourism development, shared social identity, and tourism resilience, this chapter examines three diverse Iranian heritage cases. With their different conservation approaches, these examples explain how successful conservation efforts can cultivate a stronger sense of shared identity within the community. This, in turn, fosters greater motivation and support for tourism development, ultimately contributing to a more resilient tourism industry.

1.3 Case Studies

1.3.1 Baqer Abad Castle

Baqer Abad Castle, situated in Baqer Abad Village, Bafq City, Yazd Province, stands as the largest residential castle in Iran, characterised by its unique adobe architecture set against the backdrop of the desert landscape of Yazd. Dating back approximately 200 years to the late Zandiye period, this castle is a distinguished national heritage site, showcasing the ingenuity of traditional desert architecture. Adobe, a prevalent building material in this region, consists of mud, clay, straw, and sometimes gypsum, creating thick, insulating walls ideal for withstanding desert climate extremes.

During its prime, Baqer Abad Castle served as a vital residential fortress, offering essential protection and shelter during perilous times. Unfortunately, the passage of time saw the castle succumb to neglect and decay, ultimately leading to its abandonment.

After years of abandonment, a local entrepreneur recognised the castle's potential and embarked on a transformative journey to reuse and preserve this cultural treasure adaptively. In 2014, the Cultural Heritage and Tourism Organization granted permission to local entrepreneurs to reuse the castle. This marked the beginning of a meticulous restoration and reconstruction project guided by a scientific approach that honoured the physical and social context of Baqer Abad Village (Figure 1.2).

Between late 2015 and mid-2018, extensive studies, restoration works, and equipment upgrades were undertaken at the castle. The entrepreneur's vision and commitment culminated in the castle's successful conversion to Iran's largest eco-lodge in 2018, offering accommodation for up to 120 guests. A rural anthropology museum, an ancient mill, and a traditional sauna are also other parts of this eco-lodge. The construction of this eco-lodge is specifically aimed at recreating and celebrating the traditional lifestyle of the Bafq community.

The restoration efforts and including Baqer Abad Castle in the national heritage list sparked a reconnection among residents with their historical identities tied to this landmark and its significant events. The castle's newfound status as a major tourist attraction brought visitors to the village, elevating its prominence even further, particularly on social media platforms. Following Yazd's UNESCO World Heritage designation in 2017, tourism to the region surged, with Baqer Abad Castle becoming a must-visit destination for tourists. This increase in tourism significantly boosted the local economy, benefiting village residents. In the realm of social media, the marketing team of this business (the eco-lodge) has demonstrated exceptional performance, as evidenced by the significant increase in online mentions and citations of keywords related to Bafq City, Baqer Abad Village, Baqer Abad Castle, and the Bafq Desert, particularly since 2019. This surge in online visibility reflects effective efforts to promote these attractions and evoke interest in the historical life of the Bafq community.

Figure 1.2 Baqer Abad Castle after conservation and turning into an accommodation. (Copyright: Instagram account @hana.edriss.trips)

Today, the impact of the castle's restoration and tourism development has transformed it into a major symbol of the village's history and the legacy of its ancestors. Numerous research teams dedicated their efforts to studying and restoring the castle, collaborating closely with native experts and artisans. This collective endeavour revitalised the castle and instilled a sense of pride and identity among the villagers, making tourism a resilient pillar of their community's livelihood. In other words, the castle's restoration has transformed the village into a focal point for small-scale tourism enterprises, including artisanal craft makers and other interrelated businesses. This transformation revitalised the local economy and helped bolster the resilience of the tourism ecosystem within the village.

1.3.2 Hawraman Takht

Nestled within the Zagros Mountains in the Kurdistan and Kermanshah Provinces of Iran, Hawraman Takht is the main village in the cultural landscape of Hawraman, consisting of two main valleys adapted by inhabitants over millennia to their mountainous environment. This landscape embodies the cultural heritage of the Hawrami people, a Kurdish tribe rooted in the Zagros Mountains for millennia. Their traditions—transhumance, seasonal living in Havars, steep-slope terraced stone houses, steep-slope terraced agriculture, and traditional village planning—reflect a harmonious co-existence with nature (Figure 1.3). The site is rich in intangible heritage, emphasising diverse ancestral practices that have sustained human life in this challenging terrain (UNESCO, 2021).

Despite its wealth of historical, cultural, and natural attractions, Hawraman has remained relatively unknown and has not flourished as a tourist destination for many years. The scarcity of agricultural land, limitations in

Figure 1.3 Landscape of Hawraman Takht Village. (Copyright: 3rd co-author)

mountainous livestock farming, and economic challenges have led to rural and young populations migrating to cities. Consequently, numerous valuable stone houses with unique architecture have been abandoned and left to decay. Simultaneously, traditional architectural values waned, giving way to new urban constructions using non-native and disparate materials. This approach altered the mountainous landscape, damaging pristine nature and disrupting village unity and identity by introducing generic architecture. Lack of education among locals and insufficient promotion of Hawraman's historical and cultural significance contributed to residents losing the sense of pride in their heritage and opting for modern urban transformations. This shift was evident in the decline of traditional rituals and cultural ceremonies.

In 2014, the Cultural Heritage Organization and the UNESCO World Heritage Centre in Iran decided to introduce it for inscription on the UNESCO World Heritage List. The establishment of a cultural heritage base in Hawraman and the initiation of studies to prepare the UNESCO World Heritage documentation gradually raised awareness of the tangible and intangible heritage of Hawraman. Upon discovering that their village possessed unique and significant values deserving of the UNESCO World Heritage recognition, which would attract both international and domestic tourists, people's interest in conserving their heritage and assets was significantly heightened. The awakening of a sense of identity and responsibility to protect their heritage and recognise the importance of preparing this cultural landscape for UNESCO World Heritage registration motivated the indigenous people to restore many historic stone houses voluntarily (Figure 1.4).

The inscription of the cultural landscape of Hawraman as a World Heritage Site in 2021 resulted in heritage conservation through adaptive reuse and tourism facilities development. The number of accommodations increased from one hotel and two ecolodges to several hotels and approximately 50 ecolodges by 2024, all achieved through the adaptive reuse of historic buildings transformed into local accommodations, restaurants, handicraft shops, and museums. Furthermore, this catalysed the growth of handicrafts, the founding of NGOs committed to safeguarding natural heritage, and a heightened inclination among locals to revive their traditional rituals and celebrations. These outcomes underscore the substantial positive influence of conservation initiatives and UNESCO World Heritage recognition on Hawraman's cultural identity, fostering the establishment of diverse local tourism enterprises and contributing to the resilience of local tourism in this case.

1.3.3 *Gilan Rural Heritage Museum*

The Gilan Rural Heritage Museum was established in Rasht city in 2005. Gilan Province is adjacent to the Caspian Sea amidst Hyrcanian forests and is characterised by a humid climate and terrain. Throughout history, the indigenous people of this region have employed highly creative methods to construct buildings that can withstand moisture and harmonise with the distinctive

Figure 1.4 Restoration and reuse of historical houses by local people. (Copyright: 3rd
     co-author)

climate. These unique buildings, which showcase the positive interaction
between humans and the environment, have rich historical and cultural value.
Along with the construction techniques and indigenous architectural knowledge found in the diverse buildings in this area, which have been enriched and
evolved over the years, an intangible heritage must be preserved and perpetu-
ated. Preserving these traditions is essential for maintaining cultural identity
and fostering sustainable practices that respect and harmonise with the natural world.

In 1990, a devastating earthquake measuring 7.4 on the Richter scale struck
the Gilan Province and destroyed numerous buildings and villages. Following
this catastrophic event, the Organization of Cultural Heritage of Gilan was
concerned about preserving the region's valuable historical buildings. In
response to this concern, the idea of establishing the Gilan Rural Heritage
Museum was conceived. A crucial step in constructing the museum was to
identify and acquire unique rural houses belonging to native rural communities. These houses were entrusted to specialists and experts in the restoration
and preservation of cultural heritage, who meticulously disassembled and
transported them to the museum site (Figure 1.5). Through extensive meetings
and discussions with the local people, a strong sense of cultural identity and a

Figure 1.5 (a) and (b): Reconstructed rural houses in Gilan Rural Heritage Museum. (Copyright: Instagram account @gilan_rural_heritage_museum)

desire to preserve the heritage of their ancestors emerged. This motivated them to willingly offer their houses to restoration experts to rebuild the museum site.

Furthermore, native artisans with traditional knowledge and skills played a vital role in this endeavour. They collaborated closely with the restoration team and applied their expertise to ensure that the reconstructed houses faithfully reflected the original architectural techniques and aesthetics. This collaborative effort facilitated the preservation of tangible heritage and served as a platform for safeguarding invaluable intangible cultural practices from generation to generation.

The Gilan Rural Heritage Museum, a unique institution in Asia, has reconstructed rural houses from Gilan Province, preserving its cultural traditions. This highly visited museum showcases a collection of historic buildings averaging 150 years old, aiming to display rural architecture and protect local culture and construction techniques. In addition to presenting rural architecture, the museum displays various cultural elements including tools, food, and clothing. In 2008, the museum's design received second place at the 21st Khwarizmi International Festival, a scientific and industrial festival in Iran with numerous national and international submissions, highlighting its innovative approach to heritage preservation and cultural storytelling. The museum's cultural significance has rendered it an integral component of Gilan's tourism industry, enhancing tourism resilience, particularly in terms of cultural resilience.

These conservation efforts provide both direct and indirect economic benefits for a group of residents. Residents consider it an invaluable asset, fostering a sense of identity and pride they eagerly share with visitors.

1.4 Discussion and Implications

Conserving cultural heritage is crucial, as it serves as an asset for nurturing cultural resilience in a community (Holtorf, 2018). In this regard, the primary aim of this chapter was to explore the relationships between heritage conservation, tourism development, shared social identity, and tourism resilience concepts. This exploration was facilitated by an in-depth analysis of the three distinct cases in Iran. These cases were carefully selected to highlight the variations in nature, particularly in the conservation approaches employed.

In the initial case, a historical castle was examined; it underwent restoration by the personal finances of a local entrepreneur and has since been repurposed into a tourism accommodation facility. The conservation initiative successfully preserved numerous facets of local architecture and traditions, transforming the castle into an important tourism hub in the village. This initiative has revitalised the historical site and created job opportunities for a cohort of residents, leading to significant economic benefits within the community. As a significant village symbol, the project enhanced the community's social identity and garnered their support for tourism resilience. The success of this conservation project could reinforce the region's dedication to sustainable local tourism development, which could inspire other communities.

In the second case, the conservation initiative included an entire village scale and was distinguished by the UNESCO inscription process, which heightened local community awareness regarding the intrinsic value of their identity and heritage. This awareness, in turn, stimulated voluntary conservation efforts by residents, leading to a surge in tourism-related businesses and positioning tourism as a fundamental element of the village and the community. The conservation initiative yielded favourable outcomes for the village, transforming it into a sustainable and prosperous place of interest for both tourists and residents. This case exemplifies community-based tourism resilience building through heritage conservation.

In the final case, a government expert group's conservation initiative led to preserving traditional housing styles in northern Iran, which were at risk of vanishing. Establishing a dedicated museum to house these culturally significant homes ensured the preservation of a vital piece of local heritage. This effort has created an intriguing and culturally rich complex that has become a notable tourist attraction, significantly enhancing tourism resilience in the region. The museum has become a cultural exchange and education hub, bringing locals and tourists together to appreciate the region's rich heritage. The success of this conservation initiative serves as a model for other communities facing similar challenges in preserving their cultural identity and history.

By examining the three local case studies, it can be concluded that tourism development deeply rooted in social-collective identity and shaped by heritage conservation becomes more resilient to various challenges, specifically in traditional communities where traditions and cultural roots hold profound significance for residents. Therefore, the enhancement of their identity through cultural tourism development is met with resounding support. Residents readily embrace tourism initiatives to showcase their rich cultural heritage to outsiders (domestic and international tourists), creating a deep sense of pride within the community. In other words, heritage conservation and tourism development can enhance the community's capacity to adapt to changes.

As Cartier and Taylor (2020) discuss, it is crucial to comprehend the interconnectedness between tourism development and its effects on the destination community, emphasising the community's resilience and capacity to adapt. By enhancing a community's resilience across various dimensions of community capital, such as social, physical, financial, and psychological resources, the community becomes better equipped to navigate uncertainties. This nurturing of diverse resources fosters the community's collective efforts, enabling them to effectively address various conditions and challenges (Wakil et al., 2021). Furthermore, improving the sense of pride that results in heritage conservation fosters a strong attachment to cultural assets and motivates residents to actively engage in tourism activities, thereby contributing to the sustainable growth and vitality of the local tourism sector. Generally, heritage conservation and tourism development, particularly in small societies, strengthen the shared social identity of communities and contribute to tourism resilience in countries facing economic challenges that hinder tourism development. Enhancing tourism resilience is essential to ensure future sustainability and preserve the unique features of tourism destinations.

The practical implications of this study are as follows: Promoting the community's shared identity in tourism planning and decision-making is recommended. It is imperative to cultivate collaboration among tourism stakeholders, residents, and cultural heritage authorities to guarantee that the progress of tourism is in harmony with the community's identity and values. Creating educational and cultural events to foster awareness and admiration of local heritage among residents and tourists is necessary. Additionally, it is recommended that resources be allocated towards resilience-enhancing initiatives to boost the ability of communities to adjust and react to unexpected events.

References

Bagheri, F., Hajinejad, A., & Abdi, N. (2022). Heritage conservation for tourism development: Identifying the challenges in developing countries–The case of Iran. In S. Kladou, K. Andriotis, A. Farmaki, & D. Stylidis (Eds.), *Tourism planning and development in the Middle East* (pp. 61–77). CABI.

Baglioni, M., Poggi, G., Chelazzi, D., & Baglioni, P. (2021). Advanced materials in cultural heritage conservation. *Molecules, 26*(13), Article 13. https://doi.org/10.3390/molecules26133967

Bec, A., McLennan, C., & Moyle, B. D. (2016). Community resilience to long-term tourism decline and rejuvenation: A literature review and conceptual model. *Current Issues in Tourism, 19*(5), 431–457. https://doi.org/10.1080/13683500.2015.1083538

Bui, H. T., Jones, T. E., Weaver, D. B., & Le, A. (2020). The adaptive resilience of living cultural heritage in a tourism destination. *Journal of Sustainable Tourism, 28*(7), 1022–1040. https://doi.org/10.1080/09669582.2020.1717503

Cartier, E. A., & Taylor, L. L. (2020). Living in a wildfire: The relationship between crisis management and community resilience in a tourism-based destination. *Tourism Management Perspectives, 34*, 100635. https://doi.org/10.1016/j.tmp.2020.100635

Cerisola, S., & Panzera, E. (2024). Heritage tourism and local prosperity: An empirical investigation of their controversial relationship. *Tourism Economics*, 13548166241234099. https://doi.org/10.1177/13548166241234099

Chapagain, N. K. (2016). Contextual approach to the question of authenticity in heritage management and tourism. *Journal of Heritage Management, 1*(2), 160–169. https://doi.org/10.1177/2455929616687898

Dela Santa, E., & Tiatco, S. A. (2019). Tourism, heritage and cultural performance: Developing a modality of heritage tourism. *Tourism Management Perspectives, 31*, 301–309. https://doi.org/10.1016/j.tmp.2019.06.001

Di Pietro, L., Guglielmetti Mugion, R., & Renzi, M. F. (2018). Heritage and identity: Technology, values and visitor experiences. *Journal of Heritage Tourism, 13*(2), 97–103. https://doi.org/10.1080/1743873X.2017.1384478

Erfurth, L. M., Hernandez Bark, A. S., Molenaar, C., Aydin, A. L., & Van Dick, R. (2021). "If worse comes to worst, my neighbours come first": Social identity as a collective resilience factor in areas threatened by sea floods. *SN Social Sciences, 1*(11), 272. https://doi.org/10.1007/s43545-021-00284-6

Escalera-Reyes, J. (2020). Place attachment, feeling of belonging and collective identity in socio-ecological systems: Study case of Pegalajar (Andalusia-Spain). *Sustainability, 12*(8), 3388. https://doi.org/10.3390/su12083388

Guo, Y., Zhang, J., Zhang, Y., & Zheng, C. (2018). Examining the relationship between social capital and community residents' perceived resilience in tourism destinations. *Journal of Sustainable Tourism, 26*(6), 973–986. https://doi.org/10.1080/09669582.2018.1428335

Holtorf, C. (2018). Embracing change: How cultural resilience is increased through cultural heritage. *World Archaeology, 50*(4), 639–650. https://doi.org/10.1080/00438243.2018.1510340

Jiménez-Medina, P., Artal-Tur, A., & Sánchez-Casado, N. (2021). Tourism business, place identity, sustainable development, and urban resilience: A focus on the socio-cultural dimension. *International Regional Science Review, 44*(1), 170–199. https://doi.org/10.1177/0160017620925130

Kideghesho, J. R. (2008). Co-existence between the traditional societies and wildlife in western Serengeti, Tanzania: Its relevancy in contemporary wildlife conservation efforts. *Biodiversity and Conservation, 17*(8), 1861–1881. https://doi.org/10.1007/s10531-007-9306-z

Makian, S., Borouj, A., & Hanifezadeh, F. (2022). Sustainable tourism development in rural areas: The case of community-based lodges in Iran. In S. Kladou, K. Andriotis, A. Farmaki, & D. Stylidis (Eds.), *Tourism planning and development in the Middle East* (pp. 44–60). CABI.

Mansour, H. M., Alves, F. B., & Da Costa, A. R. (2023). A comprehensive methodological approach for the assessment of urban identity. *Sustainability, 15*(18), 13350. https://doi.org/10.3390/su151813350

Mydland, L., & Grahn, W. (2012). Identifying heritage values in local communities. *International Journal of Heritage Studies, 18*(6), 564–587. https://doi.org/10.1080/13527258.2011.619554

Neville, F. G., Novelli, D., Drury, J., & Reicher, S. D. (2022). Shared social identity transforms social relations in imaginary crowds. *Group Processes & Intergroup Relations, 25*(1), 158–173. https://doi.org/10.1177/13684302220936759

Nunkoo, R., & Gursoy, D. (2012). Residents' support for tourism: An identity perspective. *Annals of Tourism Research, 39*(1), 243–268. https://doi.org/10.1016/j.annals.2011.05.006

Otero, J. (2022). Heritage conservation future: Where we stand, challenges ahead, and a paradigm shift. *Global Challenges, 6*(1), 2100084. https://doi.org/10.1002/gch2.202100084

Parga-Dans, E., González, P. A., & Enríquez, R. O. (2020). The social value of heritage: Balancing the promotion-preservation relationship in the Altamira World Heritage Site, Spain. *Journal of Destination Marketing & Management, 18*, 100499. https://doi.org/10.1016/j.jdmm.2020.100499

Ruiz Ballesteros, E., & Hernández Ramírez, M. (2007). Identity and community—Reflections on the development of mining heritage tourism in Southern Spain. *Tourism Management, 28*(3), 677–687. https://doi.org/10.1016/j.tourman.2006.03.001

Sotirova, K., Peneva, J., Ivanov, S., Doneva, R., & Dobreva, M. (2012). Digitisation of cultural heritage–standards, institutions, initiatives. In K. Ivanova, M. Dobreva, P. Stanchev, & G. Totkov (Eds.), *Access to digital cultural heritage: Innovative applications of automated metadata generation* (pp. 23–68). Plovdiv University Publishing House "Paisii Hilendarski".

Tweed, C., & Sutherland, M. (2007). Built cultural heritage and sustainable urban development. *Landscape and Urban Planning, 83*(1), 62–69. https://doi.org/10.1016/j.landurbplan.2007.05.008

UNESCO (2021). Cultural landscape of Hawraman/Uramanat. Retrieved from. https://whc.unesco.org/en/list/1647/

Wakil, M. A., Sun, Y., & Chan, E. H. (2021). Co-flourishing: Intertwining community resilience and tourism development in destination communities. *Tourism Management Perspectives, 38*, 100803. https://doi.org/10.1016/j.tmp.2021.100803

2 Mapping the Impact of COVID-19 on Tourism

A Global Bibliometric Analysis

Danijela Boljat, Daniela Garbin Praničević and Ante Mandić

2.1 Introduction

Tourism, a vital part of the global economy, was among the most severely affected industries during the COVID-19 pandemic. The crisis revealed the sector's vulnerability to unexpected disruptions, triggering widespread interest from both researchers and policymakers. Although tourism has historically shown resilience by recovering quickly from crises, the pandemic caused lasting structural changes, especially in tourist behavior and demand (Sigala, 2020). The COVID-19 pandemic thus highlighted tourism's fragility, emphasizing the need for a deeper understanding of its long-term impacts.

To analyze the pandemic's effects on tourism, bibliometric analysis offers a valuable quantitative tool. This method visualizes scientific output and allows for examining key patterns in research, including contributions from authors, institutions, and countries. Since 2020, over 5,000 papers in Scopus have addressed "COVID-19" and "tourism," exploring various dimensions of this intersection. While many studies have examined regional impacts, Uğur and Akbıyık (2020) highlight the need for more comprehensive research on global tourism's response to the pandemic.

This study applies bibliometric analysis to assess the evolving body of literature on tourism and COVID-19, identifying key trends, influential authors, and research gaps. By examining the data from 2020 to 2023, this research provides insights into the pandemic's effects on the global tourism sector and suggests areas for future inquiry. The following sections cover the research methodology, the broader implications of the pandemic on international tourism, and the results of the bibliometric analysis, concluding with key findings and recommendations.

2.2 Bibliometric Analysis

Bibliometrics is a quantitative method used to analyze scientific publications, measuring productivity, impact, and interactions across various fields (Zupic & Čater, 2015). It assesses scientific knowledge and helps identify influential papers while reducing bias. The field evolved from ancient document indexing

DOI: 10.4324/9781032720555-3

to a formal discipline in the mid-20th century, with key contributions from scholars like Cole, Eales, and Price, who advanced the study of scientific communication (Cole & Eales, 1917; Price, 1963).

Bibliometrics became more robust with the rise of computer science and bibliographic databases, marked by the launch of journals like *Scientometrics* and *Research Evaluation*. Modern bibliometrics is used by methodologists, researchers, and policymakers to analyze the development of scientific communities and the spread of ideas (Glanzel, 2003; Pehar, 2010).

This field primarily analyzes scientific articles, focusing on indicators such as citation impact and scientific collaboration. Tools like the Web of Science and metrics like the Impact Factor are essential for evaluating scientific output, reflecting a journal's influence based on citation counts (Garfield, 1979). Bibliometrics has become a key tool for researchers, institutions, and decision-makers to evaluate research quality and productivity.

2.2.1 Bibliometric Databases and Tools

Bibliometric databases vary in their suitability for different analyses, particularly in terms of scientific coverage. The three main databases frequently used are as follows:

- **Web of Science (WoS)**: Founded by Eugene Garfield in the 1960s, WoS is a subscription-based, multidisciplinary database maintained by Clarivate Analytics. It provides access to over 82 million records from 1900 to the present, covering nearly 22,000 journals, conference proceedings, and books across 256 disciplines. While comprehensive, its limitations include a subscription fee, limited social sciences and humanities coverage, and technical constraints like storing only up to 1,000 records at a time.
- **Scopus**: Introduced by Elsevier in 2004, Scopus is a multidisciplinary database that rivals WoS. It covers over 90 million records from 1970 onward, including 27,950 peer-reviewed journals and 11.6 million conference proceedings. Scopus is noted for its extensive coverage, user-friendly features, and tools for narrowing search results, virtual author identities, and citation tracking (Macan, 2007).
- **Google Scholar**: Launched in 2004, Google Scholar is a free search engine for professional and scientific literature, including peer-reviewed journals, books, conference proceedings, theses, and patents. As of 2019, it indexed around 389 million documents (Gusenbauer, 2019). However, it faces issues with data accuracy, control, transparency, and limited export capabilities, complicating large-scale analyses.

After data analysis, insights are visualized using tools that create scientific maps, illustrating connections within a field, including relationships between disciplines, specialties, papers, or authors. These maps fall into three types:

- **Citation-Based Maps**: Nodes represent publications, journals, or authors, with links indicating citations or citation-based measures like bibliographic linkage and co-citation.
- **Concept-Based Maps**: Nodes are textual elements like topics or keywords, with links showing co-occurrence frequencies.
- **Author Collaboration Maps**: Nodes represent authors, with links indicating co-authored publications.

Various tools facilitate scientific mapping, including CiteSpace (Chen, 2006), VOSviewer (McAllister et al., 2022), Sci2 Tool (Börner, 2010), Bibexcel (Persson, 2017), Biblioshiny (Aria & Cuccurullo, 2017), CitNetExplorer (Van Eck & Waltman, 2014), and SciMAT (Cobo et al., 2012).

VOSviewer, developed in 2010 by CWTS at Leiden University, is a user-friendly tool for visualizing bibliometric networks. It analyses relationships between journals, researchers, and terms, such as co-citation, bibliographic linkage, and co-authorship, and constructs co-occurrence networks using advanced text processing techniques. VOSviewer offers three display modes:

- **Network Visualization**: Shows entities like publications, journals, or researchers as nodes, with relationships like citations or co-authorship as links, helping to identify connections, clusters, and key research topics.
- **Overlay Visualization**: Uses colors to highlight specific node properties, such as citation counts, aiding in the analysis of significant nodes.
- **Density Visualization**: Displays node concentration, revealing research areas with high activity and potential collaboration hotspots.

In addition to visualizing networks, VOSviewer supports analyses like cluster analysis, centrality analysis, and temporal development analysis, offering deeper insights into scientific research (Chen, 2006). The tool is accessible online, easy to use, and customizable to meet specific research needs.

2.2.2 Bibliometric Analyses in Tourism

The growing number of bibliometric studies in tourism reflects the significant increase in publications driven by researchers' interest in various aspects of tourism, such as management, marketing, and tourist behavior. Bibliometric analysis, commonly applied to evaluate journals and authors, is widely used in tourism literature to assess productivity, citations, co-citations, keywords, authorship, leading countries, institutions, and topics (Hall, 2011).

Notable journals that have been bibliometrically analyzed include the *Journal of Sustainable Tourism, Journal of Travel & Tourism Marketing, International Journal of Hospitality Management, International Journal of Contemporary Hospitality Management*, and *Journal of Heritage Tourism*. Early studies, like McKercher's (2008), employed an "impact ratio" to identify the most influential journals. Subsequent research by Evren and Kozak (2014) explored citation trends and institutional performance in Turkish tourism literature. Other studies, such as Zopiatis et al. (2015), examined shifts in publishing strategies, highlighting increased collaboration among authors as a key development in the field. Additionally, bibliometrics can reveal spatial distribution patterns, as demonstrated by Shen et al. (2018), who created a classification of tourism and hospitality publications based on geographic location. These bibliometric analyses enhance understanding of key topics and trends in tourism research, helping scholars identify crucial elements for advancing the field.

2.3 Research Methodology

2.3.1 Data Source and Selection

For this bibliometric analysis of international tourism and the COVID-19 pandemic, Scopus was chosen as the primary database due to its extensive coverage of peer-reviewed literature, particularly in tourism and hospitality, and its compatibility with bibliometric software (Zupic & Čater, 2015). A search was conducted on May 12, 2023, using "All fields/TITLE-ABS-KEY" with keywords including "COVID-19," "tourism," and related terms. The initial query returned 1,255 publications. To refine the results, filters were applied, limiting the dataset to English-language, open-access articles published from 2020 onwards, reducing the total to 425 publications. After exporting the results to Excel, duplicates were removed, leaving 419 articles. A further screening of titles and abstracts reduced the number to 212, with a final full-text review yielding 203 relevant publications for analysis.

2.3.2 Data Analysis

The bibliometric analysis was conducted using VOSviewer version 1.6.19. This software facilitates comprehensive mapping of bibliographic data, including co-authorship, co-citation, keyword co-occurrence, citation, and bibliographic coupling. VOSviewer is particularly effective with large datasets (100+ items), offering detailed visualization and easy exploration of bibliometric maps through zoom, scroll, and search functions. Its ability to display results clearly and explicitly sets it apart from other bibliometric mapping tools (Van Eck & Waltman, 2010).

2.4 Results of Bibliometric Analysis (2020–2023)

2.4.1 Publication Trends and Global Distribution of Publications

The increasing number of publications highlights the growing research interest in the impact of COVID-19 on international tourism. Between 2020 and 2023, 203 articles were published, with a notable 133% growth in 2020, followed by a

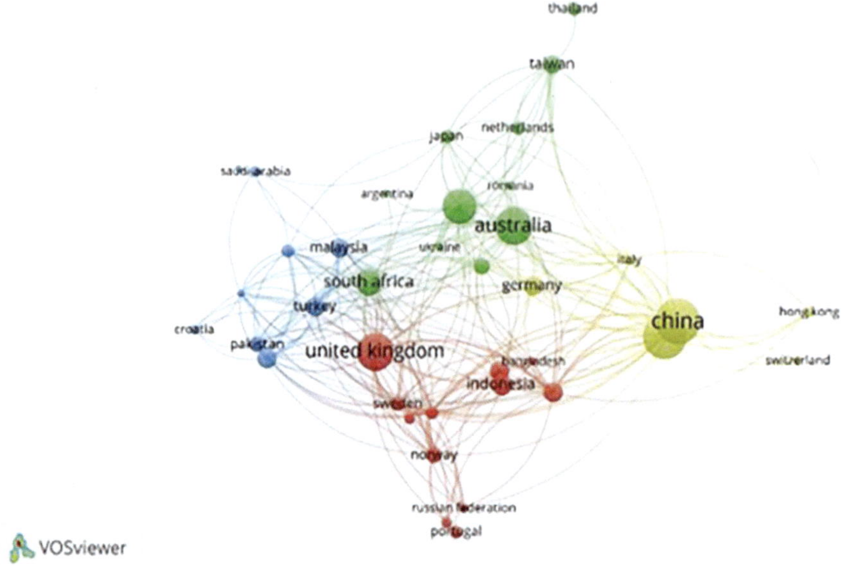

Figure 2.1 A cooperation network of countries based on research into the relationship between international tourism and the coronavirus.

Source: VOSviewer.

smaller 10% increase in 2022. A 66% decline in 2023 reflects the initial surge of research responding to the crisis (UNWTO, 2021). These publications originated from 82 countries, with China (30), the USA (29), Australia (23), the UK (23), and Spain (21) contributing the most. Economically developed countries with strong tourism sectors, like China and Spain, led the research on crisis management and recovery (UNWTO, 2021). International collaborations brought the total number of publications to 348, with China and the USA leading in global partnerships, evidenced by their high Total Link Strength (TLS) of 68, followed by the UK and South Korea (TLS 50) (Van Eck & Waltman, 2010).

In terms of citations, the UK led overall, though Canada, Sweden, and Norway had the highest average citations per publication, suggesting greater influence despite fewer publications. China and the USA ranked lower in citation impact, indicating that while prolific, their papers were less cited on average compared to smaller countries like Sweden and Norway (Figure 2.1).

2.4.2 *Leading Institutions and Authors*

Bibliometric analysis identified three universities—University of Islamabad (Pakistan), Woosong University (South Korea), and Linnaeus University (Sweden)—as the most productive, each contributing 3 publications out of 203 on international tourism and COVID-19 between 2020 and 2023. While

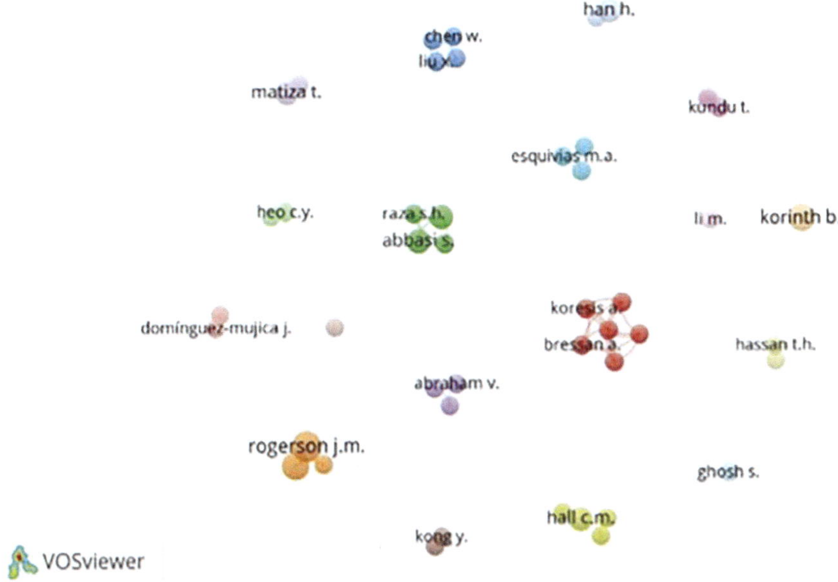

Figure 2.2 Relationship of authors and co-authors who work in the field of international tourism research and the coronavirus pandemic.

Source: VOSviewer.

publication count indicates productivity, it does not necessarily reflect research impact. Linnaeus University stands out with an exceptional 2,285 citations and an average of 761.67 citations per paper, indicating its dominant role in this research area (Scopus data).

Among the most productive authors in this field, Jayne M. Rogerson leads with five publications, though her work, along with Christian M. Rogerson's, has a lower citation impact compared to others. The most cited author is C. Michael Hall, with 2,236 citations across 3 publications, averaging 745.33 citations per paper, followed by Heesup Han (129 citations) and Tafadzwa Matiza (112 citations). This suggests that while Rogerson's are prolific, the research by Hall and others has had a more significant influence on the academic community.

The University of Johannesburg is notably represented by two of the top ten most productive authors, highlighting its strong contribution to research in international tourism and COVID-19. Overall, these institutions and authors have played a pivotal role in advancing the understanding of the pandemic's impact on global tourism. On that line, the leading institutions with more than two publications are the University of Islamabad (Pakistan), Woosong University (South Korea), and Linnaeus University in Växjö (Sweden) (Figure 2.2; Table 2.1).

Table 2.1 Authors rank according to the number of articles on international tourism and the coronavirus pandemic (authors with two or more publications)

Author	Research area / author's institution	Publications number (N = 203)	Publications distribution (%)	Citations number	Average number of citations per publication	TLS
Rogerson, Jayne M.	Economic Geography & Tourism and Sustainability/ University of Johannesburg	5	2.46	43	8.60	28
Korinth, Bartosz	Social and Economic Geography & Spatial Management/ University of Gdańsk	4	1.97	30	7.50	3
Rogerson, Christian M.	Tourism and Development/University of Johannesburg	4	1.97	29	7.25	19
Abbasi, Saba	Environment Sustainability & Management/University in Islamabad	3	1.48	35	11.67	20
Hall, C. Michael	Tourism and Human Mobility/University of Canterbury	3	1.48	2236	745.33	22
Han, Heesup	Sustainable Tourism & Hospitality Marketing/Sejong University	3	1.48	129	43.00	5
Matiza, Tafadzwa	Travel Behaviour & Tourism Brand Marketing/North-West University, Potchef stroom	3	1.48	112	37.33	5
Zaman, Umer	Tourism Marketing & Regenerative Tourism/Woosong University	3	1.48	35	11.67	20
Abraham, Villy	Consumer Behavior/Sapir College, Sderot	2	0.99	68	34.00	0
Aktan, Murat	Marketing/Muğla Sıtkı Koçman University	2	0.99	24	12.00	17

Source: Author's creation based on data from Scopus.

2.4.3 *Journal and Article Analysis*

Between 2020 and 2023, 203 articles on international tourism and COVID-19 were published across 110 journals, with *Sustainability* leading the way, publishing 32 articles (15.76%). Other notable journals include *GeoJournal of Tourism and Geosites* and the *International Journal of Environmental Research and Public Health*, each contributing ten articles. Although tourism-specific journals accounted for only 20% of publications, the interdisciplinary nature of the research highlights the pandemic's wide-ranging impact across various fields (Table 2.2).

Table 2.2 Journal rank according to the number of published articles on international tourism and the coronavirus pandemic (magazines with two or more articles)

Journal title	Number of published articles (N = 203)	Articles distribution (%)	Number of citations	Average citations per article	TLS
Sustainability	32	15.76	523	16.34	11
GeoJournal of Tourism and Geosites	10	4.93	68	6.80	14
International Journal of Environmental Research and Public Health	10	4.93	158	15.80	4
PLOS One	6	2.96	212	35.33	4
African Journal of Hospitality, Tourism and Leisure	5	2.46	26	5.20	4
Journal of Air Transport Management	5	2.46	255	51.00	2
International Journal of Hospitality Management	4	1.97	297	74.25	2
Journal of Sustainable Tourism	4	1.97	2204	551.00	6
Annals of Tourism Research	3	1.48	48	16.00	0
Journal of Tourism Futures	3	1.48	153	51.00	2
Proceedings of the National Academy of Sciences of the United States of America	3	1.48	326	108.67	5

(*Continued*)

Table 2.2 (continued)

Journal title	Number of published articles (N = 203)	Articles distribution (%)	Number of citations	Average citations per article	TLS
Scientific Reports	3	1.48	158	52.67	1
Terra Economicus	3	1.48	23	7.67	2
Tourism Review	3	1.48	511	170.33	3
Biological Conservation	2	0.99	40	20.00	1
Environment, Development, and Sustainability	2	0.99	1	0.50	1
Journal of Hospitality and Tourism Insights	2	0.99	12	6.00	2
Journal of Revenue and Pricing Management	2	0.99	37	18.50	0
Problems and Perspectives in Management	2	0.99	53	26.50	3
Quality & Quantity	2	0.99	14	7.00	0
Research in Globalization	2	0.99	22	11.00	0
Scientific African	2	0.99	3	1.50	0
Tourism and Hospitality Research	2	0.99	1	0.50	0
Tourism Management	2	0.99	25	12.50	0
Tourism Review International	2	0.99	35	17.50	5
Transport Policy	2	0.99	7	3.50	0
Transportation Research Part E: Logistics and Transportation Review	2	0.99	48	24.00	0

Source: Author's creation based on data from Scopus.

Social sciences dominated the research output, contributing 64.53% of the articles, followed by ecology and business disciplines, reflecting the complexity of COVID-19's impact on tourism. Citation metrics further emphasize the influence of key articles, with the most cited works being Gössling et al. (2020), which received 2,146 citations, and Chinazzi et al. (2020) with 2,018 citations. These foundational articles, published early in the pandemic, continue to shape the discourse on tourism and crisis management (Table 2.3).

Table 2.3 Top 3 most cited articles on international tourism and the coronavirus pandemic (with more than 50 citations)

Rank	Research paper (Author, Title, Journal, Publication year)	Citation no	Links
1.	Gössling, S., Scott, D., & Hall, C. M. (2020). Pandemics, tourism and global change: a rapid assessment of COVID-19. *Journal of sustainable tourism, 29*(1), 1–20.	2146	0
2.	Chinazzi, M., Davis, J. T., Ajelli, M., Gioannini, C., Litvinova, M., Merler, S., ... & Vespignani, A. (2020). The effect of travel restrictions on the spread of the 2019 novel coronavirus (COVID-19) outbreak. *Science, 368*(6489), 395–400.	2018	1
3.	Wen, J., Kozak, M., Yang, S., & Liu, F. (2021). COVID-19: potential effects on Chinese citizens' lifestyle and travel. *Tourism Review, 76*(1), 74–87.	412	0

Source: Author's creation based on data from Scopus.

The citation network analysis revealed three closely related clusters, with key contributions from Chinazzi et al. (2020), Gössling et al. (2020), and Duarte Alonso et al. (2020). These works have been pivotal in shaping the research landscape on the intersection of international tourism and the coronavirus pandemic (Figure 2.3).

2.4.4 Keyword Analysis

Using VOSviewer to analyze keyword co-occurrence, the most relevant keywords in research on international tourism and the coronavirus pandemic were identified to highlight current and potential future research topics. By applying a minimum occurrence filter of 5, 30 keywords were classified into 4 main groups, with the most relevant being "COVID-19," "tourism," "pandemic," and "infection control" (see Figure 2.4).

- **Group 1 (Red)**: Focuses on the impact of COVID-19 on the tourism sector, particularly in terms of future development, recovery, and management. This group addresses how tourism can adapt to new challenges and pursue sustainable development.
- **Group 2 (Green)**: Centers on the connection between COVID-19 infections and the measures taken, such as quarantine and travel restrictions.
- **Group 3 (Light Blue)**: Examines the pandemic's economic impact on the tourism sector.
- **Group 4 (Yellow)**: Investigates changes in travel behavior due to the pandemic and the implementation of travel restrictions to curb the virus's spread. Understanding these dynamics is crucial for informed tourism management and planning safe, sustainable travel.

Figure 2.3 Relationship between publications and citations for publications cited at least 50 times.

Source: VOSviewer.

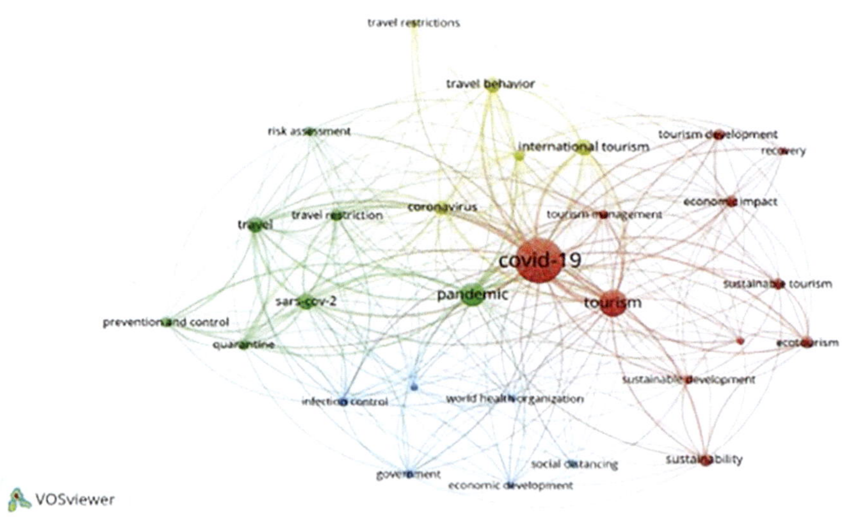

Figure 2.4 Network visualization of co-occurrence of keywords in the Scopus database.

Source: VOSviewer.

The analysis identified the top 30 keywords, with "COVID-19" occurring 161 times and "tourism" 59 times, followed by "pandemic" with 45 occurrences. These keywords reflect the primary research areas and emerging themes within the intersection of tourism and the pandemic. This keyword analysis helps define specific research topics, guiding future studies and contributing to the ongoing discourse on managing the impacts of COVID-19 on global tourism.

2.5 Discussion and Conclusion

The keyword analysis of literature on COVID-19 and international tourism reveals several critical trends that underscore both the challenges and opportunities brought by the pandemic. While the pandemic undeniably dealt a severe blow to the tourism industry, it also catalyzed a rethinking of how the industry operates, emphasizing the urgent need for restructuring toward sustainability. This shift is evident in the frequent appearance of sustainability-related themes in the literature. Scholars like Seabra and Bhatt (2022) and Chowdhury and Jomo (2020) argue that the pandemic offers a unique opportunity to reassess and improve upon pre-pandemic practices, steering the industry toward more sustainable and responsible tourism. This emphasis on sustainability is not just a temporary reaction but also a strategic pivot that could define the future trajectory of global tourism.

The research also highlights the growing importance of rural, community-based, and local tourism as strategies to enhance the industry's resilience to future crises. The pandemic has demonstrated that these forms of tourism can play a crucial role in mitigating the impacts of global disruptions, thereby contributing to a more robust and adaptable tourism sector. This shift toward localized tourism aligns with broader sustainability goals, suggesting that the future of tourism may increasingly prioritize local experiences and community involvement.

Sustainability frameworks have been applied extensively to analyze various pandemic-related challenges, such as reducing greenhouse gas emissions (Gössling & Higham, 2020) and mitigating the pressure on natural ecosystems (Bates et al., 2021). Additionally, the pandemic has spurred interest in alternative tourism models, such as virtual travel, which, while nascent, offer new avenues for sustainable tourism practices (Spenceley et al., 2021; Talwar et al., 2022). This research contributes to understanding how the tourism industry can evolve to meet new environmental and social challenges, making it more resilient and sustainable in the long term.

The analysis of pandemic-related measures, such as social distancing, quarantines, vaccinations, and digital COVID certificates, is critical for understanding how tourism destinations and companies have adapted to new realities (Chinazzi et al., 2020; de Miguel Beriain & Rueda, 2022; Meng et al., 2021; Radic et al., 2021; Sotis et al., 2021; Zaman et al., 2022). These adaptations are not only reactive but also offer valuable lessons for future crisis management in the tourism sector. By exploring these measures, the research provides insights into the practical aspects of managing tourism during a global health crisis, contributing to the development of more effective and adaptable strategies for future challenges.

The pandemic's impact on tourist behavior and perceived risks has been another focal point of research. Understanding how tourists' perceptions and behaviors have shifted is essential for the tourism industry to adapt its offerings and communication strategies (Abraham et al., 2020; Bremser et al., 2022; Matiza, 2022; Matiza & Slabbert, 2022; Rasoolimanesh et al., 2021; Wu et al., 2023). This research contributes to a more nuanced understanding of these changes, offering valuable insights that can help the industry better align with the evolving expectations and concerns of travelers.

The early research on COVID-19's impact on international tourism was largely descriptive and conceptual, laying the groundwork for more in-depth studies. While this initial research was crucial in framing the challenges, there is now a clear need for more advanced studies that offer actionable strategies and solutions. This study identifies gaps in the literature, particularly the need for more detailed analyses of specific strategies and policies that can drive sustainability in tourism post-pandemic. Furthermore, the research highlights the importance of developing practical approaches for destinations and tourism businesses to adapt to new conditions and attract tourists in a post-pandemic world.

The period from 2020 to 2023 witnessed intense scientific activity, with 203 articles from 82 countries analyzed, reflecting the global importance of this research topic. The dramatic increase in publications in 2020, with a 133% surge, underscores the immediate and widespread interest in understanding the pandemic's impact on tourism. The leading contributions from countries like China, the USA, Australia, the UK, and Spain indicate that economically developed nations are at the forefront of researching this global issue, likely due to their significant stakes in the international tourism industry.

The bibliometric analysis also identified key academic institutions contributing to this field, with the University of Islamabad, Woosong University in South Korea, and Linnaeus University in Sweden emerging as leaders. Linnaeus University, in particular, stands out for its high citation count, indicating the substantial influence of its research on the broader academic community. The prominence of the journal *Sustainability* as the preferred publication venue further emphasizes the centrality of sustainability in ongoing tourism research.

This study's multidisciplinary approach is another critical contribution, as it reflects the complex, interconnected nature of the pandemic's impact on tourism. The involvement of researchers from diverse fields such as social sciences, ecology, and business economics underscores the importance of integrating multiple perspectives to fully understand and address the challenges posed by the pandemic. Finally, the research highlights the pivotal role played by seminal articles, such as those by Gössling et al. and Chinazzi et al., in shaping the early discourse on tourism and COVID-19. These foundational studies have not only provided critical insights but also set the agenda for future research, guiding the development of strategies to navigate the post-pandemic landscape of global tourism. In summary, this research offers valuable contributions to the understanding of the pandemic's impact on international tourism. It underscores the importance of sustainability, highlights the shift in tourist behavior, and provides insights into the strategies needed to adapt to and recover from such global disruptions. Despite some limitations, such as the focus on Scopus-indexed, English-language articles published until May 2023, this study offers a comprehensive overview that can guide future research and policymaking in the tourism sector.

2.5.1 Study Limitations and Future Research

This study has several limitations. It focuses only on articles indexed in the Scopus database, excluding potentially relevant studies from other sources and non-English publications. The analysis is limited to publications up to May 2023, missing more recent developments. Additionally, the reliance on bibliometric analysis may overlook qualitative insights. Future research should expand to include more databases, languages, and recent publications. Incorporating qualitative methods could also provide deeper insights into the pandemic's impact on tourism, informing more comprehensive recovery and sustainability strategies.

References

Abraham, V., Bremser, K., Carreno, M., Crowley-Cyr, L., & Moreno, M. (2020). Exploring the consequences of COVID-19 on tourist behaviors: Perceived travel risk, animosity and intentions to travel. *Tourism Review, 76*(4), 701–717. https://doi.org/10.1108/TR-07-2020-0344

Alonso, A. D., Kok, S. K., Bressan, A., O'Shea, M., Sakellarios, N., Koresis, A., ... & Santoni, L. J. (2020). COVID-19, aftermath, impacts, and hospitality firms: An international perspective. *International Journal of Hospitality Management, 91*, 102654. https://doi.org/10.1016/j.ijhm.2020.102654

Aria, M., & Cuccurullo, C. (2017). A brief introduction to bibliometrix. *Journal of Informetrics, 11*(4), 959–975.

Bates, A. E., Primack, R. B., Biggar, B. S., Bird, T. J., Clinton, M. E., Command, R. J., Richards, C., Shellard, M., & Vergara, V. (2021). Global COVID-19 lockdown highlights humans as both threats and custodians of the environment. *Biological Conservation, 263*(109175). https://doi.org/10.1016/j.biocon.2021.109175

Börner, K. (2010). *Mapping scientific networks*. Microsoft PowerPoint - 10-VIVO-Conf [Compatibility Mode]

Bremser, K., Crowley-Cyr, L., Abraham, V., Moreno-Martin, M. J., & Carreño, M. (2022). Application of the health belief model to explain public perceptions, travel intentions and actions during COVID-19: A sequential transformative design. *Journal of Hospitality and Tourism Insights, 5*(5), 865–885.

Chen, C. (2006). CiteSpace II: Detecting and visualizing emerging trends and transient patterns in scientific literature. *Journal of the American Society for information Science and Technology, 57*(3), 359–377.

Chinazzi, M., Davis, J. T., Ajelli, M., Gioannini, C., Litvinova, M., Merler, S., Pastore, A., Rossi, L., Sun, K., Viboud, C., Xiong, X., Yu, H., Halloram, E., Longini Jr., I. M., Vespignani, A. (2020). The effect of travel restrictions on the spread of the 2019 novel coronavirus (COVID-19) outbreak. *Science 368*,395–400. https://doi.org/10.1101/2020.02.09.20021261

Chowdhury, A. Z., & Jomo, K. S. (2020). Responding to the COVID-19 pandemic in developing countries: Lessons from selected countries of the global south. *Development, 63*(2), 162–171.

Cobo, M., López-Herrera, A., Herrera-Viedma, E., & Herrera, F. (2012). SciMAT: A new science mapping analysis software tool. *Journal of the American Society for Information Science and Technology, 63*, 1609–1630. https://doi.org/10.1002/asi.22688

Cole, F. J., & Eales, N. B. (1917). The history of comparative anatomy: Part I.—A statistical analysis of the literature. *Science Progress (1916–1919), 11*(44), 578–596.

De Miguel Beriain, I., & Rueda, J. (2022). Immunity passports, fundamental rights and public health hazards: A reply to Brown et al. *Journal of Medical Ethics, 46*, 660–661. https://doi.org/10.1136/medethics-2020-106814

Evren, S., & Kozak, N. (2014). Bibliometric analysis of tourism and hospitality related articles published in Turkey. *Anatolia, 25*(1), 61–80. https://doi.org/10.1080/13032917.2013.824906

Garfield, E. (1979). Scientometrics comes of age. *Current Contents*, (46), 5–10.

Glanzel, W. (2003). *Bibliometrics as a research field a course on theory and application of bibliometric indicators* (course book). https://schola+r.google.com/schola+r?hl=en&as_sdt=0%2C5&q=Glanzel%2C+W.+%282003%29.+Bibliometrics+as+a+resea+rch+field+a+course+on+theory+and+application+of+bibliometric+indicators+&btnG=#d=gs_cit&t=1733843216416&u=%2Fschola+r%3Fq%3Dinfo%3A3g9NU1c448EJ%3Aschola+r.google.com%2F%26output%3Dcite%26scirp%3D0%26hl%3Den

Gössling, S. & Higham, J. (2020). The low carbon imperative: Destination management under urgent climate change. *Journal of Travel Research*. https://doi.org/10.1177/0047287520933679

Gössling, S., Scott, D., & Hall, C. M. (2020). Pandemics, tourism and global change: A rapid assessment of COVID-19. *Journal of Sustainable Tourism, 29*(1), 1–20. https://doi.org/10.1080/09669582.2020.1758708

Gusenbauer, M. (2019). Google Scholar to overshadow them all? Comparing the sizes of 12 academic search engines and bibliographic databases. *Scientometrics, 118*, 177–214. https://doi.org/10.1007/s11192-018-2958-5

Hall, C. M. (2011). Publish and perish? Bibliometric analysis, journal ranking and the assessment of research quality in tourism. *Tourism Management, 32*(1), 16–27. https://www.e-unwto.org/doi/pdf/10.18111/9789284422456

Macan, B. (2007). Scopus – nova generacija baza podataka. *Kemija u industriji, 56*(2), 64–68. https://hrcak.srce.hr/12342

Matiza, D. T. (2022). Country image and recreational tourism travel motivation: The mediating effect of south Africa's place brand dimensions. *Tourism and Hospitality Management, 28*(3), 519–539. https://doi.org/10.20867/thm.28.3.2

Matiza, T., & Slabbert, E. (2022). The destination media profile and tourist travel intentions: The mediating effect of Covid-19 induced perceived risk. *Advances in Hospitality and Tourism Research (AHTR), 10*(4). https://doi.org/10.30519/ahtr.943356

McAllister, J. T., Lennertz, L., & Atencio Mojica, Z. (2022). Mapping a discipline: A guide to using VOSviewer for bibliometric and visual analysis. *Science & Technology Libraries, 41*(3), 319–348.

McKercher, B. (2008). A citation analysis of tourism scholars. *Tourism Management, 29*(6), 1226–1232.

Meng, F., Gong, W., Liang, J., Li, X., Zeng, Y., & Yang, L. (2021). Impact of different control policies for COVID-19 outbreak on the air transportation industry: A comparison between China, the U.S. and Singapore. *PLoS One, 16*(3). https://doi.org/10.1371/journal.pone.0248361

Pehar, F. (2010). Od statističke bibliografije do bibliometrije. Povijest razvoja kvantitativnog pristupa istraživanju pisane riječi. *Libellarium, 3*(1), 1–28. https://hrcak.srce.hr/74289

Persson, O. (2017). *Bibexcel. A tool-box programme for bibliometric analysis.* https://homepage.univie.ac.at/juan.gorraiz/bibexcel

Price, D. J. D. S. (1963). *Little science, big science.* Columbia University Press.

Radic, A., Koo, B., Gil-Cordeno, E., Cabrera-Sánchez, J. P., & Han, H. (2021). Intention to take covid-19 vaccine as a precondition for international travel: Application of extended norm-activation model. *International Journal of Environmental Research and Public Health, 8*(6), 1–15. https://doi.org/10.3390/ijerph18063104

Rasoolimanesh, S. M., Seyfi, S., Rastegar, R., & Hall, C. M. (2021). Destination image during the COVID-19 pandemic and future travel behavior: The moderating role of past experience. *Journal of Destination Marketing and Management, 21*(1). https://doi.org/10.1016/j.jdmm.2021.100620

Seabra, C., & Bhatt, K. (2022). Tourism sustainability and COVID-19 pandemic: Is there a positive side? *Sustainability, 14*, 8723.

Shen, Y., Morrison, A. M., Wu, B., Park, J., Li, C., & Li, M. (2018). Where in the world? A geographic analysis of a decade of research in tourism, hospitality, and leisure journals. *Journal of Hospitality & Tourism Research, 42*(2), 171–200. https://doi.org/10.1177/1096348014563394

Sigala, M. (2020). Tourism and COVID-19: Impacts and implications for advancing and resetting industry and research. *Journal of Business Research, 117*, 312–321. https://doi.org/10.1016/j.jbusres.2020.06.015

Sotis, C., Allena, M., Reyes, R., & Romano, A. (2021). Covid-19 vaccine passport and international traveling: The combined effect of two nudges on Americans' support for the pass. *International Journal of Environmental Research and Public Health, 18*(162). https://doi.org/10.3390/ijerph18168800.

Spenceley, A., Mccool, S., Newsome, D., & Báez, A. (2021). Tourism in protected and conserved areas amid the COVID-19 pandemic. *PARKS. The International Journal of Protected Areas and Conservation, 27*(Special Issue), 103–118. https://doi.org/10.2305/IUCN.CH.2021.PARKS-27-SIAS.en

Talwar, S., Kaur, P., Nunkoo, R., Dhir, A.(2022). Digitalization and sustainability: Virtual reality tourism in a post pandemic world. *Journal of Sustainable Tourism.* https://doi.org/10.1080/09669582.2022.2029870

Uğur, N. G., & Akbıyık, A. (2020). Impacts of COVID-19 on global tourism industry: A cross-regional comparison. *Tourism Management Perspectives, 36*. https://doi.org/10.1016/j.tmp.2020.100744

Van Eck, N., & Waltman, L. (2010). Software survey: VOSviewer, a computer program for bibliometric mapping. *Scientometrics, 84*(2), 523–538. https://doi.org/10.1007/s11192-009-0146-3

Van Eck, N. J., & Waltman, L. (2014). CitNetExplorer: A new software tool for analyzing and visualizing citation networks. *Journal of Informetrics, 8*(4), 802–823.

World Tourism Organisation [UNWTO]. (2021). UNWTO world tourism barometer. https://webunwto.s3.eu-west-1.amazonaws.com/s3fs-public/2021-10/UNWTO_Barom21_05_September_excerpt.pdf?tObUi1QiC40DQFbrkfyryClQWEF3KFf7

Wu, C. K., Ho, M. T., Trag Le, T. K., & Nguyen, M. U. (2023). The COVID-19 pandemic and factors influencing the destination choice of international visitors to Vietnam. *Sustainability, 15*(1), 396. https://doi.org/10.3390/su15010396

Zaman, U., Koo, I., Abbasi, S., Raza, S. H., & Qureshi, M. G. (2022). Meet your digital twin in space? Profiling International expat's readiness for metaverse space travel, tech-savviness, COVID-19 travel anxiety, and travel fear of missing out. *Sustainability*, *14*(11). https://doi.org/10.3390/su14116441

Zopiatis, A., Theocharous, A. L., & Constanti, P. (2015). The past is prologue to the future: An introspective view of hospitality and tourism research. *Scientometrics*, *102*, 1731–1753. https://doi.org/10.1007/s11192-014-1431-3

Zupic, I., & Čater, T. (2015). Bibliometric methods in management and organization. *Organizational Research Methods, 18*(3), 429–472.

3 Destination Resilience through Stakeholder Collaboration in Destination Crisis Management

Christin Khardani

3.1 Introduction

Since the outbreak of the Covid-19 pandemic, the terms 'crisis' and 'resilience' have become frequently used buzzwords in the public and academic discourse in connection with tourism. Tourists represent a particularly vulnerable group of people in a destination (Berbekova et al., 2021) while in turn, the tourism destination itself is prone to external crises (Filimonau & Coteau, 2020) such as natural disasters ranging from tropical storms to earthquakes and floods, epidemics such as Ebola or the recent Covid-19 pandemic, terrorist attacks, political unrest like the so-called Arabic Spring or financial crises. Another particularity of tourism destinations is their composition of multiple suppliers that operate under different conditions in different businesses (e.g. tourism industry, hospitality and retail) and sectors (e.g. public or private) on different spatial levels (e.g. regional, national, or international) (Pavlovich, 2003). To date, the application of stakeholder collaboration theories is still limited in tourism crisis management (Jiang & Ritchie, 2017). This is why academics call for an approach to how to overcome the disconnect between parties during a crisis (Morakabati et al., 2017). Stakeholder collaboration is also considered a fundamental part of resilience building in destinations, which allows them to successfully cope with change (Filimonau & Coteau, 2020). Before engaging in the topic of stakeholder collaboration, it is necessary to take a step back and first determine the actors that shape the tourism destination (Wang et al., 2022).

Outbound tour operators play a vital role in a destination during a crisis. They decide about the handling of the guests and whether to offer the destination in the future. However, there is only little academic research about their role in crisis management (Sausmarez, 2013). The role of DMOs has been widely discussed in academic literature. Lately, there has been an outcall for new tasks, particularly with respect to crisis handling before, during and after a crisis (e.g. Basurto-Cedeño & Pennington-Gray, 2016; Borzyszkowski, 2013).

This chapter seeks to illuminate the current state of collaboration among selected tourism stakeholders in crisis management by addressing the research

DOI: 10.4324/9781032720555-4

question: 'To what extent is the collaboration between tourism stakeholders articulated in the crisis management documents of national and international associations?' Moreover, this chapter contributes to research on this topic by providing a status-quo analysis after discussing the peculiarities of crisis management in destinations and its link to destination resilience. The chapter closes with practical implications for destinations to build resilience.

3.2 Crisis Management in Destinations – Same but Different Approach

The term 'crisis' has been defined in various ways without reaching a consensus. According to Faulkner (2001), a crisis is a self-inflicted event occurring from within an organisation while a disaster is an external event that can neither be avoided nor controlled by the organisation. Some researchers follow this differentiation (e.g. Runyan, 2006; Xu & Grunewald, 2009) while others use 'crisis' and 'disaster' synonymously. Although crises in a destination are usually externally induced, this chapter follows the latter approach and applies the rather broad definition of Pauchant and Mitroff (1992, p. 15), who define a crisis as a 'disruption that physically affects a system as a whole and threatens its basic assumptions, its subjective sense of self, its existential core'.

Crisis management comprises all the tasks that capacitate an organisation to successfully handle a crisis which includes mitigation, preparedness, response and recovery (Fink, 1986). The substantial difference to strategic management is that decisions need to be made under the following conditions:

- lack of (clear) information,
- immense time pressure,
- pressure from the public,

while

- the degree of control is low and
- response options are few (Burnett, 1998)

Those challenges are already difficult to handle within a single organisation. The complexity of destination crisis management might become clear here: It is not one organisation that must handle a crisis but multiple ones that operate within a complex system – the tourism destination (Wang et al., 2022). It consists of numerous private and public actors within and outside the tourism industry (Pavlovich, 2003), which are highly dependent on each other (Berbekova et al., 2021) and are obligated to collaborate to successfully deal with a crisis (Mistilis & Sheldon, 2006). A second main challenge of crisis management in destinations is the tragedy of commons (Hardin, 1968) as the main resources that define a destination are usually common goods (Zemla, 2016). Those include the following resources:

- physical, e.g. beaches, lakes, national parks,
- built, e.g. museums or heritage locations,
- intangible, e.g. destination brand, reputation of local society's attitude towards life (Zemla, 2016).

As a result, important questions that need to be answered before a crisis hits the destination are:

- Who should take over the ownership of those common goods during a crisis?
- Who coordinates the interaction between all stakeholders involved?
- Who is responsible for the long-term recovery until the destination reaches its status quo and can go 'back to normal' or even 'build back better'. In short: Who is responsible for building a resilient destination? The impact of individual organisations might be limited, but at the same time, their contribution is vital to support the recovery process of the destination (Xu & Grunewald, 2009).

One could assume that the acting agent here should be the DMOs. Before discussing the key stakeholders in the crisis management of destinations, a brief discourse on resilience in the context of crisis management in destinations is provided to give both concepts a framework.

3.3 The Contribution of Crisis Management to Destination Resilience

Holling (1973, p. 14) was the first to introduce the concept of resilience within the ecologic field as the ability of systems 'to absorb change and disturbance and still maintain the same relationships between populations or state variables'. Since then, resilience has been applied across disciplines such as medicine, psychology, economics, engineering, social sciences and – since the beginning of the 21st century – also in tourism (Brown et al., 2017). Among scholars, the desirable outcome of resilience can be divided into three streams:

- A group of scholars, usually with engineering backgrounds, considers resilience as a system's capacity to resist a disturbance (Demiroz & Haase, 2019).
- Some consider resilience as a strategy to recover from an external shock with the aim of rebuilding the status quo of the system (Basurto-Cedeño & Pennington-Gray, 2016).
- Others understand resilience as an opportunity for change and a driver for innovation after the disruption of the system (Cahyanto & Pennington-Gray, 2017).

Destinations should mainly focus on the third approach since in the course of natural disasters boosted by climate change, destinations are forced to reinvent themselves.

Depending on the research field, there exist various types of resilience such as organisational, economic, ecological, engineering, evolutionary, community or systems resilience (Brown et al., 2017; J. Cheer & Lew, 2018). Organisational resilience is the organisation's ability to detect, manage and adapt to vulnerabilities within a complex and dynamic environment (McManus et al., 2008). Tourism stakeholders must first build organisational resilience to enable the tourism destination to achieve resilience as a system through stakeholder collaboration (Cochrane, 2010).

Resilience and crisis management are both dynamic concepts and mutually dependent: Destination crisis management can be regarded as the path to destination resilience as it supports the destination to successfully deal with crises by providing mitigation, preparation, response and recovery strategies (e.g. Brown et al., 2017; Cochrane, 2010; Wang et al., 2022). Vice versa, resilience can also be seen as a proactive approach towards crisis management because a resilient destination is prepared for any occurring crisis and ready to deal with and adapt to it (Beirman, 2018). A disaster-resilient destination contributes to a sustainable destination and community (Beirman, 2018; Brown et al., 2017). Therefore, scholars propose to merge the approaches of sustainable tourism and tourism resilience by adding the dimension of resilience to the triangle of sustainability currently embracing economic, ecological and social aspects (Cochrane, 2010; Filimonau & Coteau, 2020).

3.4 Untangling Stakeholders in a Destination During Crisis

Stakeholders that contribute to the tourism product in a destination are the hospitality industry offering accommodation and catering, tourism companies such as Destination Management Companies (also called incoming agencies), tour guides, transportation services or operators of tourist attractions as well as government agencies usually represented by the DMOs (also called national tourism authority or organisation) (Zemla, 2016). The product created in the destination is either combined with a package vacation and sold by outbound tour operators and/or (online) travel agencies in the respective source market or booked separately and autonomously by the tourists. Cooperation among these stakeholders is already complex and dynamic in everyday business (Zemla, 2016). During a crisis, the pressure for and among those stakeholders increases while cooperation becomes even more essential for the destination's survival. In their analysis of 15 crisis management models with tourism focus Khardani and Schmude (2024) reveal that tourism stakeholders are only marginally broken down into detail as most of the models refer broadly to 'tourism businesses' and 'DMOs' only while tour operators are not mentioned at all.

However, tour operators can have a major influence on a destination because, with their package holidays, they offer the most influential product within the destination, especially in Europe (Picazo & Moreno-Gil, 2018). This gives them particular bargaining power in the destination in terms of rates and allotment contracts and turns them into an influential stakeholder in classic package holiday destinations, which are usually the so-called sun and beach destinations (Picazo & Moreno-Gil, 2018). With this market power, large tour operators are able to shape whole destinations because they also have an influence on the transport connections (e.g. flight slots) of a destination. Vertically integrated tourism companies have a financial stake in destinations where they build their own tourism product offering accommodation, transfer and tour leader services from a single source. They can develop and at the same time close a destination for international tourists, thus destinations can be heavily dependent on them. A current example of the negative consequences of such large tourism companies is the insolvency of the third largest European tour operator FTI Group in June 2024 leaving more than 10,000 employees in hotels and DMOs in the destinations without a job (Jacobs, 2024).

The previously discussed complexity of a tourism destination needs to be managed, which is usually the task of the DMOs (Roach et al., 2023). DMOs are run by the government, non-profit organisations or private institutions (Basurto-Cedeño & Pennington-Gray, 2016) and operate depending on the geographical and political definition of the destination on regional, national or international levels (Roach et al., 2023). Until recently, their tasks mainly concentrated on marketing, sales and information for consumers and business partners (Hartman et al., 2020). In terms of crisis management, DMOs gradually gained importance in the recovery phase of crisis management by communicating the restored destination to consumers and by negotiating marketing incentives with business partners (Borzyszkowski, 2013). DMOs interact with public institutions (e.g. governments and local authorities), non-governmental organisations (NGOs, e.g. local community) and private companies (e.g. touristic service providers) (Becken, 2013) with whom they need to build a strong partnership to ensure the collaboration needed following a crisis (Jiang & Ritchie, 2017). At the same time, they are also expected to coordinate and meet the requirements of all involved actors during a crisis (Roach et al., 2023). Additionally, tourists are particularly vulnerable to crisis as they do not know the environment and may encounter language barriers (Faulkner, 2001). Therefore, the DMOs are considered to be their main contact for support during and after a crisis (Basurto-Cedeño & Pennington-Gray, 2016).

3.5 The Underestimated Role of Tour Operators and DMOs in Crisis Management – A Status-Quo Analysis

The relevance of tour operators and DMOs in regional, national and international crisis plans published by national organisations or international bodies was analysed using document analysis. These crisis plans emerged during internet research using the terms 'crisis plans' AND 'crisis management plans' AND 'disaster plans' AND 'disaster management plans' to identify the generic crisis plans. The same procedure was repeated with the addition of 'tourism' AND to identify crisis plans with tourism focus. The document research was completed by conducting a snowball approach using the references of the identified documents. Research in scientific databases was not conducted because the research focuses specifically on non-scientific publications. The following criteria were chosen for the analysis:

- Spatial level
- Approach towards crisis management
- Publishing institution
- Year of publication
- Target group
- 'Collaboration' or 'cooperation' mentioned
- 'Tour operator' mentioned (in tourism documents only)
- 'DMO' or 'Destination Management Organisation' mentioned (in tourism documents only)

In total, 20 documents were identified with regional (9), national (3) and international (8) spatial levels published either by national organisations (e.g. DRV and PATA) or by international bodies (e.g. UNEP and WTTC) both with a broad (10) and a touristic approach (10) towards crisis management. Table 3.1 gives a detailed picture of the analysed elements 1–4.

Among the 15 identified institutions that offer crisis or disaster management plans, the 'Caribbean Disaster Emergency Management Agency' (CDEMA) is the institution with the most publications (4) followed by the 'United Nations' (UN) in different sub-sections (3). The 'Asia-Pacific Economic Cooperation' (APEC) provides 2 documents while the remaining 12 institutions issue a single document each.

All documents have been published between 2006 and 2022. This aligns with the academic research on tourism crisis management, which has increased since the terrorist attacks in the United States of America in September 2001 (Jiang et al., 2019). During the mentioned period, one or two documents were published yearly except for four years where no publications appeared.

Figure 3.1 shows that most of the crisis plans are dedicated to governments or authorities followed by the tourism industry. The latter is addressed more precisely in some documents. The stakeholders of the tourism industry like tour operators, DMOs or tourism destinations are named as target groups in two documents each. This also applies to other stakeholders like businesses, NGOs and emergency response teams. Crisis plans are an integral part of crisis management. It is therefore surprising that other stakeholders, but governments, are addressed so little.

Within the generic crisis plans, the words 'collaboration' or 'coordination' are mentioned between 0 and 46 times and among the tourism crisis plans between 0 and 26 times. Overall, the distribution varies strongly. Cautiously interpreted, those results show that collaboration does not seem to have a great significance in crisis plans.

Looking at Figure 3.2, tour operators are mentioned 178 times in total in 80% of the documents with an uneven distribution. The German Travel Association DRV stands out by mentioning tour operators 72 times. The document by COMCEC mentions tour operators 54 times, while it appears

Table 3.1 Selected criteria of crisis plan of national or international associations sorted by spatial level (n = 20)

Spatial level	Approach towards crisis management	Publishing institution	Document name	Year of publication
International	Generic	United Nations Development Program	Methodological guide for post-disaster recovery planning processes	2011
International	Generic	United Nations Educational, Scientific and Cultural Organization	Managing disaster risks for world heritage	2010
International	Generic	Capacity for disaster reduction initiative	Basics of capacity development for disaster risk reduction	2011
International	Generic	Northwest Healthcare Response Network	Guidelines for writing an emergency preparedness plan	2018
International	Generic	Global Facility for Disaster Risk Reduction and Recovery	Post-disaster needs assessments volume A guidelines	2013
International	Touristic	United Nations Environment Program	Disaster risk management for coastal tourism destinations responding to climate change. A practical guide for decision-makers.	2008
International	Touristic	Global Rescue and World Travel & Tourism Council	Crisis readiness. Are you prepared and resilient to safeguard your people & destinations?	2019
International	Touristic	The World Bank Group	Expecting the unexpected: tools and policy considerations to support the recovery and resilience of the tourism sector	2022

(*Continued*)

Table 3.1 (Continued)

Spatial level	Approach towards crisis management	Publishing institution	Document name	Year of publication
National	Generic	Government of the Commonwealth of Dominica	Post-Disaster Needs Assessment Hurricane Maria September 18, 2017. A Report by the Government of the Commonwealth of Dominica.	2017
National	Touristic	Council of Australian Tour Operators	Tourism Risk, Crisis and Recovery Management Guide	2016
National	Touristic	German Travel Association	DRV crisis guideline for tour operators	2008
Regional	Generic	Caribbean Disaster Emergency Management Agency	Regional infrastructure for information sharing and development and adaptation of model contingency ICT plan	2010
Regional	Generic	Caribbean Disaster Emergency Management Agency	Standard for: Exercise planning, execution & evaluation, part 2. Tsunami and earthquake drills	2015
Regional	Generic	Caribbean Disaster Emergency Management Agency	Comprehensive Disaster Management: A model national CDM policy: Adaptation Guide	2012
Regional	Generic	Caribbean Disaster Emergency Management Agency	Comprehensive Disaster Management: A model national CDM policy for Caribbean Countries	2012
Regional	Touristic	Caribbean Tourism Organization	Multi-hazard risk management guide for the Caribbean tourism sector	2020

(*Continued*)

Table 3.1 (Continued)

Spatial level	Approach towards crisis management	Publishing institution	Document name	Year of publication
Regional	Touristic	Asia-Pacific Economic Cooperation	How to develop a risk management strategy for a tourism business/ organisation. Participant's workbook.	2006
Regional	Touristic	Asia-Pacific Economic Cooperation	Tourism Risk Management for the Asia Pacific Region: An authoritative guide for managing crises and disasters	2006
Regional	Touristic	Pacific Asia Travel Association	Crisis Communication Planner	2020
Regional	Touristic	Standing Committee for Economic and Commercial Cooperation of the Organization of Islamic Cooperation	Risk & Crisis Management in Tourism Sector: Recovery From Crisis in the OIC Member Countries	2017

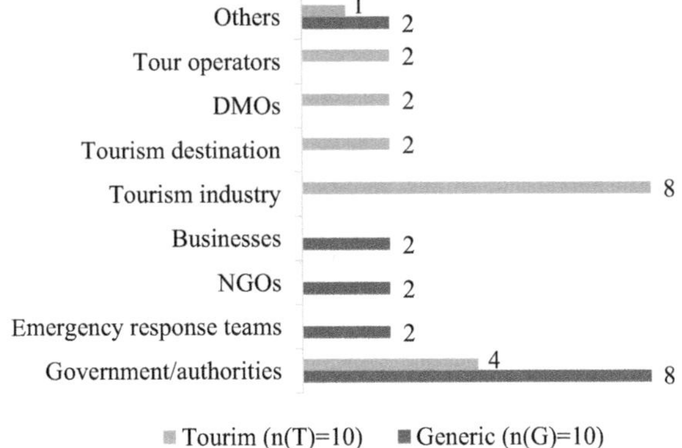

Figure 3.1 Target groups of analysed crisis plans with generic and tourism approach in total numbers (n = 20; multiple assignments).

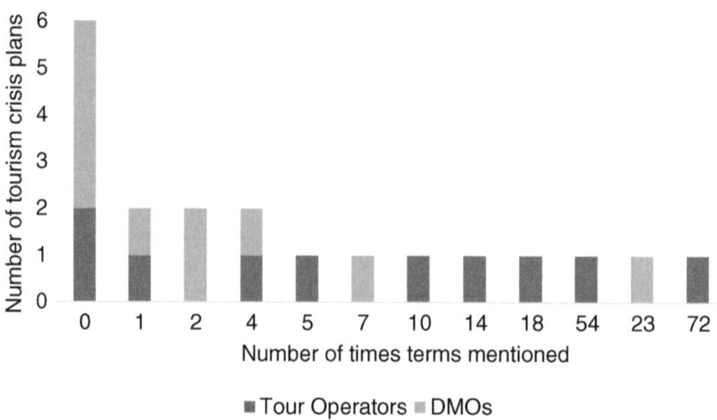

Figure 3.2 'DMO/Destination Management Organisation' and 'Tour operator' mentioned in analysed crisis plans with tourism approach in total numbers (n = 10).

18 times in the crisis plan of the Council of Australian Tour Operators (CATO). Figure 3.2 also shows that DMOs do not seem to play a major role in the tourism crisis plans as they are not mentioned at all in 40% of the documents. In total, the term appears 39 times while again, the COMCEC's document stands out with a reference of 23 times.

The frequent mention in the crisis plans published by tour operators is evident. Nevertheless, the results are surprising as they stand in direct contradiction to the academic work: An analysis of crisis management models in tourism reveals that tour operators are not mentioned at all while DMOs are mentioned in 46% of the models (Khardani & Schmude, 2024). This shows a significant difference in the perception between practitioners and academics. On the other hand, the review of academic literature conducted for this chapter clearly elaborates the stake that DMOs have in destination crisis management, which is not reflected in the document analysis of the crisis management plans. Again, theoretical implications and practical implementation do not align. The next section aims to bridge this gap by giving hints for resilience building in destinations.

3.6 Future Recommendations for Resilience Building in Destinations

Based on the findings discussed before, three actions are proposed to build destination resilience towards crises. Crisis management as part of strategic planning in the destination. Before striving for resilience, it is fundamental that tourism destinations integrate crisis management as an integral part of their strategic planning, which is increasingly internationally recognised (Beirman, 2018). It is therefore a precondition that each entity within the destination is prepared for a crisis (Mair et al., 2016). This not only reduces the level of uncertainty for each stakeholder but also benefits the whole destination (Berbekova et al., 2021).

Identify stakeholders within and outside the destination. Consequently, to build resilience, it is necessary to identify the stakeholders in the destination. In this context, it is important to consider not only the own destination but also the tourists' source markets, including the tourism service providers such as tour operators and the local authorities there (Mäntyniemi, 2012). Tour operators should be involved as influencing partners, as they have a strong stake in the destinations' economic progress not only during and after but also before a crisis. Other destinations, which are either geographically located nearby or show similar characteristics, may also have a significant stake (Beritelli, 2011).

Define and strengthen the role of the DMOs in a destination. The role of the tourism industry needs to be strengthened within the destination that emphasises the importance of the DMOs as connecting link between private companies and public agencies. Those diverse agents with their specific expertise need to work together under stressful conditions in a crisis. In other words: Those agencies acting as independent entities before the crisis need to temporarily form a supra-organisation during a crisis (e.g. Curnin et al., 2015; Hartman et al., 2020). The DMOs should therefore act as coordinators, networkers and leaders during a crisis, which requires skilled and experienced staff (Beritelli, 2011). It is not only excepted to do so by other stakeholders (Beirman, 2018), but

- The DMOs usually have all necessary contacts of public and private institutions and NGOs within and outside the destination,
- they have the necessary reputation and image to coordinate,
- in contrast to other stakeholders involved, they have or should not have a conflict of interest as they strive to decide whatever is best for the destination.

Figure 3.3 takes up the complexity of those proposals by drafting the involved stakeholders in a crisis. Within the destination, these are the public entities

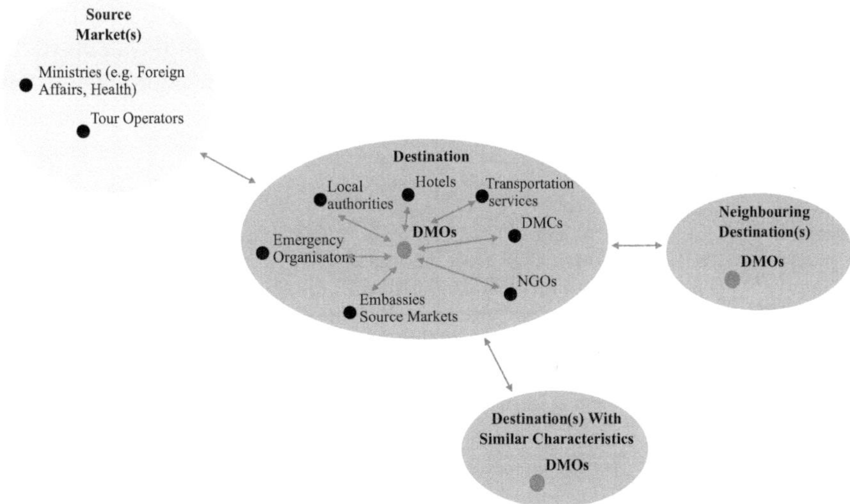

Figure 3.3 Destination resilience through stakeholder collaboration.

(local authorities, emergency organisations, source market embassies/consulates, NGOs), tourism service providers (hotels, DMCs, transportation services) and the DMOs as coordinators. Stakeholders outside the destination are the destination's main source markets with the responsible ministries and the tour operators as main actors. Additionally, possible stakeholders could be neighbouring destinations or destinations with similar characteristics. It therefore becomes clear that a leading institution needs to be assigned.

3.7　Conclusion

Berbekova et al. (2021, p. 1) express very clearly that 'in today's reality, crisis management is no longer an extra, but rather a principal and crucial function for tourism destinations and organizations'. This fact must first be internalised by all actors involved before taking the next step towards destination resilience, which should be targeted by all tourism destinations as it offers insurance for their future existence.

This chapter clearly emphasises that cooperation is a key element of successful crisis management, specifically in tourism destinations that need to meet various interests of tourists, tourism companies, governments and communities, both in the source market and the destination.

The role of tour operators and DMOs was discussed in this context. The analysis of crisis plans of national and international associations shows that those two stakeholders are taken up to varying degrees. Generally, collaboration does not seem to play a major role in those documents. This lack of stakeholder collaboration was also discovered by Filimonau and Coteau (2020), although researchers agree that a crisis can be handled successfully only when all stakeholders involved are prepared and collaborate (Mair et al., 2016).

Therefore, destinations are strongly advised to consider crisis management as part of their strategic planning. They need to be aware of all stakeholders both within and outside the destination that might be involved in a crisis. Tour operators from key source markets should not be ignored in this context. To put those approaches into practice, DMOs should take the coordinating role for crisis management in a destination.

References

Basurto-Cedeño, E. M., & Pennington-Gray, L. (2016). Tourism disaster resilience scorecard for destinations (TDRSD): The case of Manta, Ecuador. *International Journal of Tourism Cities*, 2(2), 149–163. https://doi.org/10.1108/IJTC-01-2016-0002

Becken, S. (2013). Developing a framework for assessing resilience of tourism subsystems to climatic factors. *Annals of Tourism Research, 43*, 506–528. https://doi.org/10.1016/j.annals.2013.06.002

Beirman, D. (2018). Thailand's approach to destination resilience: An historical perspective of tourism resilience from 2002 to 2018. *Tourism Review International, 22*(3), 277–292. https://doi.org/10.3727/154427218X15369305779083

Berbekova, A., Uysal, M., & Assaf, A. G. (2021). A thematic analysis of crisis management in tourism: A theoretical perspective. *Tourism Management, 86*, 104342. https://doi.org/10.1016/j.tourman.2021.104342

Beritelli, P. (2011). Cooperation among prominent actors in a tourist destination. *Annals of Tourism Research, 38*(2), 607–629. https://doi.org/10.1016/j.annals.2010.11.015

Borzyszkowski, J. (2013). Destination management organizations (DMOs) and crisis management. *Journal of Tourism and Services, 4*(7), 4–15.

Brown, N. A., Rovins, J. E., Feldmann-Jensen, S., Orchiston, C., & Johnston, D. (2017). Exploring disaster resilience within the hotel sector: A systematic review of literature. *International Journal of Disaster Risk Reduction, 22*, 362–370. https://doi.org/10.1016/j.ijdrr.2017.02.005

Burnett, J. J. (1998). A strategic approach to managing crises. *Public Relations Review, 24*(4), 475–488. https://doi.org/10.1016/S0363-8111(99)80112-X

Cahyanto, I., & Pennington-Gray, L. (2017). Toward a comprehensive destination crisis resilience framework. In *Travel and Tourism Research Association: Advancing Tourism Research Globally (2017 ttra International Conference)*. https://core.ac.uk/download/pdf/84289275.pdf

Cheer, J., & Lew, A. A. (2018). Understanding tourism resilience: Adapting to social, political, and economic change. In J. M. Cheer & A. A. Lew (Eds.), *Routledge advances in tourism. Tourism, resilience and sustainability: Adapting to social, political and economic change* (pp. 3–17). Routledge.

Cochrane, J. (2010). The sphere of tourism resilience. *Tourism Recreation Research, 35*(2), 173–285. https://doi.org/10.1080/02508281.2010.11081632

Curnin, S., Owen, C., Paton, D., & Brooks, B. (2015). A theoretical framework for negotiating the path of emergency management multi-agency coordination. *Applied Ergonomics, 47*, 300–307. https://doi.org/10.1016/j.apergo.2014.10.014

Demiroz, F., & Haase, T. W. (2019). The concept of resilience: A bibliometric analysis of the emergency and disaster management literature. *Local Government Studies, 45*(3), 308–327. https://doi.org/10.1080/03003930.2018.1541796

Faulkner, B. (2001). Towards a framework for tourism disaster management. *Tourism Management, 22*(2), 135–147. https://doi.org/10.1016/S0261-5177(00)00048-0

Filimonau, V., & Coteau, D. de (2020). Tourism resilience in the context of integrated destination and disaster management (DM 2). *International Journal of Tourism Research, 22*(2), 202–222. https://doi.org/10.1002/jtr.2329

Fink, S. (1986). *Crisis management: Planning for the inevitable.* Amacom American Management Association.

Hardin, G. (1968). The tragedy of the commons. *Science (New York, N.Y.), 162*(3859), 1243–1248. https://doi.org/10.1126/science.162.3859.1243

Hartman, S., Wielenga, B., & Heslinga, J. H. (2020). The future of tourism destination management: Building productive coalitions of actor networks for complex destination development. *Journal of Tourism Futures, 6*(3), 213–218. https://doi.org/10.1108/JTF-11-2019-0123

Holling, C. S. (1973). Resilience and stability of ecological systems. *Annual Review of Ecology and Systematics, 4*(1), 1–23. https://doi.org/10.1146/annurev.es.04.110173.000245

Jacobs, H. (2024, July 9). FTI insolvency at a glance. *Fvw.* https://www.fvw.de/international/travel-news/the-latest-news-at-a-glance-what-happens-after-the-insolvency-of-fti-touristik-243517

Jiang, Y., & Ritchie, B. W. (2017). Disaster collaboration in tourism: Motives, impediments and success factors. *Journal of Hospitality and Tourism Management, 31*(4), 70–82. https://doi.org/10.1016/j.jhtm.2016.09.004

Jiang, Y., Ritchie, B. W., & Benckendorff, P. (2019). Bibliometric visualisation: An application in tourism crisis and disaster management research. *Current Issues in Tourism, 22*(16), 1925–1957. https://doi.org/10.1080/13683500.2017.1408574

Khardani, C., & Schmude, J. (2024). A micro-level model for crisis management in tourism destinations – An interdisciplinary approach. *Journal of Contingencies and Crisis Management, 32*(3), Article e12619. https://doi.org/10.1111/1468-5973.12619

Mair, J., Ritchie, B. W., & Walters, G. (2016). Towards a research agenda for post-disaster and post-crisis recovery strategies for tourist destinations: A narrative review. *Current Issues in Tourism, 19*(1), 1–26. https://doi.org/10.1080/13683500.2014.932758

Mäntyniemi, P. (2012). An analysis of seismic risk from a tourism point of view. *Disasters, 36*(3), 465–476. https://doi.org/10.1111/j.1467-7717.2011.01266.x

McManus, S., Seville, E., Vargo, J., & Brunsdon, D. (2008). Facilitated process for improving organizational resilience. *Natural Hazards Review, 9*(2), 81–90. https://doi.org/10.1061/(ASCE)1527-6988(2008)9:2(81)

Mistilis, N., & Sheldon, P. (2006). Knowledge management for tourism crises and disasters. *Tourism Review International, 10*(1), 39–46. https://doi.org/10.3727/154427206779307330

Morakabati, Y., Page, S. J., & Fletcher, J. (2017). Emergency management and tourism stakeholder responses to crises: A global survey. *Journal of Travel Research, 56*(3), 299–316. https://doi.org/10.1177/0047287516641516

Pauchant, T. C., & Mitroff, I. I. (1992). *Transforming the crisis-prone organization: Preventing individual, organizational and environmental tragedies* (1st ed.). The Jossey-Bass management series. Jossey-Bass. https://permalink.obvsg.at/AC00682130

Pavlovich, K. (2003). The evolution and transformation of a tourism destination network: The Waitomo Caves, New Zealand. *Tourism Management, 24*(2), 203–216. https://doi.org/10.1016/S0261-5177(02)00056-0

Picazo, P., & Moreno-Gil, S. (2018). Tour operators' marketing strategies and their impact on prices of sun and beach package holidays. *Journal of Hospitality and Tourism Management, 35*, 17–28. https://doi.org/10.1016/j.jhtm.2018.02.004

Roach, C. M. L., Lewis-Cameron, A., & Brown-Williams, T. (2023). Organizational resilience in public sector organizations: Destination management organizations. *Journal of Health and Human Services Administration, 46*(1), 26–49. https://doi.org/10.37808/jhhsa.46.1.2

Runyan, R. C. (2006). Small business in the face of crisis: Identifying barriers to recovery from a natural disaster. *Journal of Contingencies and Crisis Management, 14*(1), 12–26. https://doi.org/10.1111/j.1468-5973.2006.00477.x

Sausmarez, N. de (2013). Challenges to Kenyan tourism since 2008: Crisis management from the Kenyan tour operator perspective. *Current Issues in Tourism, 16*(7–8), 792–809. https://doi.org/10.1080/13683500.2013.785488

Wang, T., Yang, Z., Chen, X., & Han, F. (2022). Bibliometric analysis and literature review of tourism destination resilience research. *International Journal of Environmental Research and Public Health, 19*(9). https://doi.org/10.3390/ijerph19095562

Xu, J., & Grunewald, A. (2009). What have we learned? A critical review of tourism disaster management. *Journal of China Tourism Research, 5*(1), 102–130. https://doi.org/10.1080/19388160802711444

Zemla, M. (2016). Tourism destination: The networking approach. *Moravian Geographical Reports, 24*(4), 2–14. https://doi.org/10.1515/mgr-2016-0018

4 Tour guides, risk and trust

An analysis of trustworthiness in tour guides during travel disruptions

*Conor McTiernan, Ciara Quinlan,
Ciarán Ó hAnnracháin and
Aylin Poroy Arsoy*

4.1 Introduction

Tourism researchers increasingly note the impact of trust on tourists' behaviours and perceptions of satisfaction (Nunkoo & Gursoy, 2016; Zhong et al., 2021). As tourism necessitates displacement from prime residencies, the tourist may face travel disruptions resulting in unfamiliarity and ambiguity, which, for some tourists, constitutes a risk (Williams & Baláž, 2020). Such risks are subjective, but they can manifest an unease due to a lack of tacit knowledge of other places (Boksberger & Craig-Smith, 2006) and can trigger concerns relating to personal safety, health or security (Nunkoo et al., 2012). Here the tourist is obliged to adopt a crisis management perspective to minimise the impact of the disruption on the overall purpose of travel. For example, the recent Covid-19 pandemic and the ever-changing rules for domestic, inbound and outbound travel heightened threats of personal health risks, often due to causal ambiguity (Williams et al., 2022). To understand how such threats can be averted, academic research suggests that the development of trust between the tourist and key stakeholders can reverse such risks or uncertainties (O'Malley et al., 2022), as trust exists where certainty ends (Park, 2020).

Given the potential for travel disruption risks, for some tourists, the engagement of a tour guide (TG) for a part or the complete duration of a visit can facilitate the reduction of threats or uncertainties (Huang et al., 2010). Equally, it must be remembered that tourists from different countries may have dissimilar perceptions of safety in terms of socioeconomic, environmental, societal, political and other potential risk factors (Seabra et al., 2013), and this requires the TG to be able to moderate the management of travel disruption and safety accordingly. By considering such risks, TGs can empathise with tourists and develop action-orientated solutions to address each individual risk factor (Patwardhan et al., 2020), and such actions encourage the development of trust between the TG and the tourists (Pagliara et al., 2021). The challenge for tourism scholars is to understand how such trust is initiated and how it develops over time. Research suggests that trust is formed on the basis of the perceived trustworthiness of the TG, where trustworthiness is founded on

DOI: 10.4324/9781032720555-5

recognised indicators of trust (Kramer, 2009). The purpose of the chapter is to explore the key indicators of trustworthiness in TGs as perceived by tourists. To begin, this research introduces the theme of crisis management, explores the role of TGs in tourism and continues by examining how trust between the TG and tourists can mitigate against risk.

4.2 Crisis management and travel disruptions

The management of risk, disaster and crisis management emanates from organisational research primarily focused on exploring crisis outcomes and the associated stakeholders' perceptions of organisational reputation, trust and legitimacy (Bundy et al., 2017). Crisis management in tourism is critical to ensure the resilience and sustainability of the industry in the face of various unforeseen events. As travel disruptions constitute both internal and external organisational crises, effective crisis management strategies are essential for minimising the negative impacts on destinations, businesses and travellers while facilitating swift recovery (Zhong et al., 2021). Therefore, crisis management in tourism requires proactive planning and crisis response research regularly draws on attribution theory (AT), where individuals are motivated to search for the causes of negative or unexpected events (Çakar, 2020). Here it is incumbent on destinations and tourism enterprises to anticipate potential crises and develop comprehensive contingency plans to address various scenarios. These plans should include protocols for communication, evacuation, resource allocation and coordination with relevant authorities and stakeholders (Le Roux & Van Niekerk, 2019).

As communication is paramount during a crisis, AT is a helpful framework for evaluating the relationship between a crisis and the selection of communication strategies (Jackson, 2019). Timely and accurate information dissemination is essential to ensure the safety of tourists and locals alike and mitigate against misinformation and panic. Stakeholders can establish communication channels through various mediums to keep tourists informed about the situation, safety measures and any changes in travel advisories (Ţuclea et al., 2020). By fostering strong partnerships between the tourist and tourism stakeholders, destinations can leverage collective expertise and resources to address challenges more effectively and expedite recovery efforts (Morakabati et al., 2017). This research proposes that the TG can act as the interlocutor between tourists and relevant stakeholders, and the tourists' perceptions of trust in the TG impact evaluation of response strategy effectiveness (Brown et al., 2016).

4.2.1 *Tour guides*

The TG can be defined as the person, typically a professional, who guides individuals and groups around natural and man-made venues or places of interest, interpreting the cultural and natural heritage in an engaging and informed means (Black & Ham, 2005). For many tourists, the primary task of the TG is as an experience broker, based on their skills, communication and interpretation

(Weiler & Black, 2015), yet such a simplistic perspective fails to appreciate the range of responsibilities bestowed upon the TG. The TG acts as the primary contact for the customer, and their perceived tacit knowledge of the destination provides confidence to customers, especially when they seek advice and recommendations (Cheng et al., 2017). TGs are often employees of tourism product organisations or destination marketing organisations, entities ultimately liable for non-performance or improper performance of tourism services (Cavlek, 2002), such as travel delays or cancellations. It is therefore incumbent on both TGs and other tourism stakeholders to develop an understanding of tourists' perceptions of risks and threats, and by addressing these, the TG can be considered trustworthy.

As tourists' perceptions of risk and safety resulting from travel disruptions can vary, research contests that tourist-perceived safety can be addressed by considering the suitability of the facilities, equipment and supporting infrastructure in the destination where the disruption has occurred (Kapuściński & Richards, 2016). Further considerations include being conscious of tourism and non-tourism employees behaviours (Peccei & Rosenthal, 2000), adhering to tourism safety management guidelines (Xie et al., 2021) and protecting the natural elements of the contextual ecosystem and social environment (Okuyama, 2018). This suggests that a key role of the TG during disruptions is to conduct, and be seen to conduct, appropriate safety and risk assessments, even more critically than an individual would.

To mitigate against travel crises, the TGs can decide on a number of strategies including whether or not to include specific locations on an itinerary, to reduce activities at a particular destination or to take immediate measures to protect their clients if required (Cavlek, 2002). By completing a safety analysis and choosing an appropriate course of action, the TG is evidencing that they are trustworthy and prioritising the needs of the tourists. Having explored how the presence of trust between tourists and TG reduces risks, the following section elaborates on the importance of trust between actors in a tourism context.

4.2.2 *Trust*

It is argued that a universal definition for trust within a tourism context has yet to be established, though some tourism researchers have tried (McTiernan et al., 2023). Tourism researchers typically define trust in terms of beliefs and expectations relating to the behaviours and intentions of others, often emphasising the vulnerabilities of the actors (Williams & Baláž, 2020). For example, scholars suggest that trust implies a willingness to be vulnerable to the actions of another (Mayer et al., 1995) or a willingness to rely on another party (Pagliara et al., 2021). Other scholars perceive trust as a choice behaviour, where actors allow themselves to become more vulnerable over time (van der Werff & Buckley, 2017). Such acceptance of vulnerability is not without risk, yet Rousseau et al. (1998) assert that trust is an acceptance of vulnerability based on positive expectations. The inclusion of positive expectations is important as it introduces a subjective probability that the trustee (TG) is motivated to meet the expectations of the trustor (tourist) (Nunkoo & Gursoy, 2016) and is trustworthy in terms of their benevolence and ability to meet expectations (Kumar et al., 2020). Importantly, this supports the contention that trust is a measure of trustworthiness (Colquitt et al., 2011).

This interpretation points to an emerging challenge in trust scholarship: the assessment of the dimensional nature of trust. Fundamentally, the dimensions of trust impact the lens through which the operationalisation of trust is assessed (Legood et al., 2022). Recent studies suggest two distinct paradigms of thought have emerged (Tomlinson et al., 2020). Firstly, trust can be conceptualised in two dimensions: cognitive and affective trust (Johnson & Grayson, 2005). The second paradigm conceptualises trust as distinct from trustworthiness and is informed by the trustee's perceptions of the ability, benevolence and integrity of the trusted actor (Mayer et al., 1995). Therefore, this research considers the indicators of trust and trustworthiness from both historic and contemporary research.

For example, early researchers posited that *integrity, motives, consistency of behaviour, openness, discreetness, interpersonal* and *functional competence, predictability* and *sound judgement* are key dimensions of trust that influence trustees (Nooteboom, 2022). Butler (1991) developed a 'conditions of trust inventory' to include such terms as *loyalty, accessibility, availability, competence, consistency, fairness, integrity, loyalty, openness, overall trust, promise fulfilment* and *receptivity*. Sankowska (2013) also identified similar attributes of trustworthiness as *competence, reliability* and *concern*. While all such terms directly impact contractual relationships during travel disruptions, *concern* is important for our TG-tourist dynamic as it can be construed as a judgement of empathy from the perspective of both the trustee and trustor.

These conditions of trust in play between the tourist and the TG are as important to this study as the context where such trust is formed and must consider the notions of rules and fairness (Hourigan, 2015). Such fairness attempts to reduce uncertainty in the TG-tourist relationship and protect against episodic or systematic exploitation of power. We submit the fundamental balance of power lies primarily with the TG, as the tourists lack appropriate tacit knowledge (Williams and Baláž, 2020), which significantly informs the contextual impact of each of the dimensions of trust. This suggests that the tourist is motivated to trust the TG during travel crises, especially when the tourist perceives that the TG possesses trustworthy traits. This research supports the view that trust is dynamic, and therefore the following sections explore the importance of examining the bases of trust within specific contexts, primarily based on the ability, integrity and benevolence of the TG (Mayer et al., 1995).

4.2.3 Ability

Contemporary quantitative and qualitative trust researchers maintain that Mayer et al.'s (1995) ability, integrity and benevolence (ABI) are essential variables when assessing the development of trust in interpersonal and inter-organisational trust studies. Ability refers to the skills, competencies and characteristics of the party to complete a task (Mayer et al., 1995), in this case, the ability of the TG to manage the travel disruption. Other academics have used similar constructs such as *competence* or *perceived expertise* (Czernek & Czakon, 2016). The ability-based trust of an actor suggests that the trustee will do what they said they would do (Nooteboom, 2022), and while this explanation is simplistic, we suggest there are two complex, interlocking dynamics in play: expertise and willingness. Expertise denotes the accumulation of sufficient tacit knowledge in a skill or process, such as tour guiding or managing travel disruptions, which is not easily achieved and clearly identifies the knowledge source as competent or skilled in the activity (Connelly et al., 2018). The second element, willingness, refers to the readiness or degree of preparedness of the TG to take action (Rousseau et al., 1998). Interestingly, researchers agree in their classification of ability-based trust as cognitive (Mayer et al., 1995) and Colquitt et al. (2011) noted its use as a tool to operationalise cognition-based trust. Ability-based trust suggests that the TG manages the travel disruption by adhering to an itinerary, keeping to an agreed schedule and meeting budgets. As ability-based trust is cognitive in nature, the tourist will be able to experience or evidence the actions and behaviours of the TG during the tour that encourage, or discourage, the formation of trust such as the development of good safety measures, which, in turn, plays an important role in building competence-based trust (Sarfraz et al., 2022).

4.2.4 Perceived competence of the TG

While ability-based trust infers trustworthiness in the TG grounded in evidence, researchers have long acknowledged the importance of a tourists' presumptive trust in the TG (van der Werff & Buckley, 2017). One such form of presumptive trust is *role-based trust*, where the trust is built on the knowledge that a particular person occupies a role rather than a specific knowledge of that person's capabilities or intentions (Kramer, 2009). Such trust is believed to be bestowed to the trustee based on their knowledge, education or period of time in that occupation, and often it's not the person that is trusted, but the system of expertise that created the role-appropriate behaviours of the occupants of a specific role (Meyerson et al., 2006). For example, the tourist may assume that the TG is being truthful when delivering information. Equally, scholars note that this form of trust is considered fragile and can result in catastrophic failures of cooperation (Nunkoo et al., 2012).

Rule-based trust also informs a conscious assessment of a scenario based on shared understandings of the systems of rules that govern appropriate behaviours (Kramer, 2009). It is suggested that rule-based trust is created and maintained not through explicit contracts but rather through socialisation of the rules (Czernek & Czakon, 2016). Tourism studies contend that when confidence in the system and its rules is high, then a trustor will have high levels of trust or even confer a taken-for-granted level of trust (Sparks & Browning, 2011). For example, if a tourist deems the stakeholders in a destination to be competent overall, rule-based trust suggests that TGs at the destination will also be competent and capable of managing travel disruptions.

This is similar to *category-based trust*, where trust is based on the trustee's membership in a social or organisational category, and often the trustor is unaware that judgements of the trustee's trustworthiness have been made (Kramer, 2009). For some researchers, category-based trust is linked to cultural stereotypes (Meyerson et al., 2006), where discerning categories can be determined through such trivial factors as the TGs uniform (Robert et al., 2009). In this context, it could be suggested that tourists are positively disposed to trusting the TG strictly on the basis that the individual is deemed qualified by their organisation and therefore adheres to an international standard expected of TGs. Cumulatively, role, rule and category-based trust add to the tourist's perception of promise fulfilment.

4.2.5 *Promise fulfilment*

Promise fulfilment refers to trust in actors where perceived value fulfilments have been achieved in the past, and the trustor believes the actor will be trustworthy in the future (Kramer, 2009). We postulate two key indicators of perceived value fulfilment in this context. Firstly, we speculate that the TG has been employed following recommendations from others or research of customer reviews. Such reviews constitute what is termed in the trust literature as third-party references as proxies of trust (Czernek & Czakon, 2016). The second indicator of promise fulfilment is the tourists' post-consumption perceptions of the value of money they experienced (Choi et al., 2018).

Kramer (2009) suggests that it is reasonable to support third-party proxies of trust, and these may form the base for presumptive trust in others. Third parties such as TGs can act as conduits of trust for the tourist as they relate trust-relevant information through exchanges, such as sharing updates on travel delays or cancellations. Such third-party references could be obtained through online research (Han et al., 2022) or through conversations with peers at, or away from, the destination (Chang, 2014). It can also be suggested that in such circumstances, TGs as third parties may only relate that which they believe the recipient wishes to hear (Huang et al., 2010), which may be a narrow perspective. Ultimately, TGs may act as important go-betweens, and based on a recommendation from a trusted party, a degree of trust is implied by the trustor (Tourist) on the trustee (TG) (Czernek & Czakon, 2016).

4.2.6 *Benevolence-based trust*

A further base of relational trust refers to trust developed over time between trustor and trustee through repeated interactions, exchanges and risk-taking, namely benevolence-based trust (Rousseau et al., 1998). Such trust indicates the extent to which the trustee believes the trustor fundamentally wants to do good (Chang, 2014). Similar terms include *intentions, motives, altruism* and *loyalty* (Xie et al., 2021). This form of trust is supported by other researchers using differing terms, for example, 'affective trust' (McAllister, 1995).

Given that TG-tourist exchanges during travel disruptions involve human interaction for the purpose of sharing a variety of forms of knowledge and that the process will be influenced by personal, organisational and structural influences, every successful tour requires an appropriate level of relational trust (Chang, 2014). The strength of the social constructs within the TG-tourist relationship acts as an enabler of trust development between the actors, which in turn may act as a counterbalance to risks perceived during potential environmental uncertainty (Williams & Baláž, 2020), such as travel delays. Benevolence-based trust creates social capital that is embedded in the relationships between people and developed through regular social interaction (Blumberg et al., 2015). It is a complex phenomenon that allows both the TG and the tourist to understand the social environment through emotional, situational and cognitive dimensions (van der Werff & Buckley, 2017). This inclusion of both cognitive and affective elements is important, as social capital development necessitates both physical and emotional triggers, often focusing on the perception that the TG is consciously ensuring the vulnerabilities of the tourists are not being exploited during travel disruptions.

4.2.7 *Perceived integrity of TG*

Mayer et al. (1995) contend that as trust is a psychological state, parties must experience or sense that partners are abiding by similar guiding principles, such as integrity-based trust, and such principles are deemed acceptable by the tourist as the trustor. Integrity-based trust has long been recognised as evidence of moral character and a desirable trait in perceived leaders, which summons relational capital through organisational commitment and obligation to the group (Chang, 2014). In return, integrity-based trust ensures sustainable rewards, reduces organisational exposure to exploitation and, most importantly, gives psychological reassurance that the intentions and behaviours of others are favourable. In this context, the integrity-based trust is not viewed as either calculative or relational; rather, it is a 'climate' that determines adherence to pre-ordained norms between TG and tourists' (Han et al., 2022). This research speculates that social capital can be strengthened between TGs and tourists, based upon adherence to the principles of integrity, and this facilitates a powerful affective-trust bond between the tourist and the TG. Where both TG and tourist share a commonality of purpose and vision, coupled with integrity-based trust, the overall tourist experience is enhanced.

In summary, the model in Figure 4.1 proposes five determinants of trust-based consumer satisfaction in TG experiences during crises or travel disruptions.

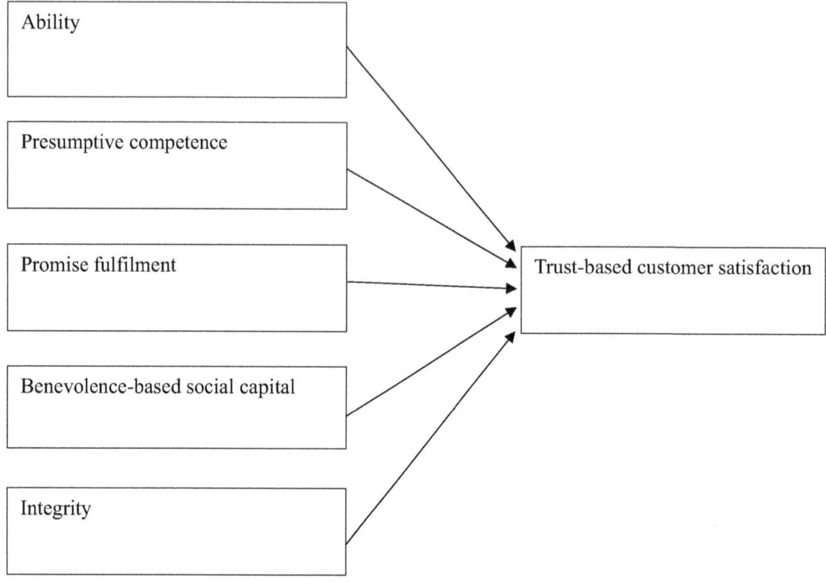

Figure 4.1 Determinants of trust-based consumer satisfaction in tour guide experiences.

4.3 Conclusion and further research

Travel disruptions by their nature create uncertainty and present risks due to a lack of tacit knowledge, which can force the tourist to adopt a crisis management position. This research considers the role of TGs as brokers of influence and contends that the physical and psychological strain of travel disruptions can be alleviated by encouraging the development of trust in the TG-tourist relationship. By drawing on such fields as tourism, psychology and crisis management, to explore how trust intervenes in the management of disruptions, the research returns to fundamental conceptualisations of trust and its implications in a specific tourism context. This chapter argues that there are not only rational elements of trust at play; the TG-tourist relationship is fundamentally influenced by both cognitive and affective indicators of trustworthiness. As trust is a psychological perception, tourists must see and feel that the risks posed by the disruption are understood by the TG, who empathise with the tourist.

While this study contributes to the emerging field of trust research in tourism, further research is required to empirically test the validity of the conceptual model. There are gaps in the knowledge determining the extent of influence these indicators have on trust development, if any. Importantly, such studies could assess the weighted importance of each indicator based on the idiosyncrasies of different travel disruptions and the resulting operational constraints on the TG. Finally, the chapter acknowledges that perceptions of risk, trust and crisis are personal and can be influenced by demographic and psychographic factors. Further studies could use a single travel disruption to explore how trust development differs between multiple tourists and TGs experiencing a single disruption.

References

Black, R., & Ham, S. (2005). Improving the quality of tour guiding: Towards a model for tour guide certification. *Journal of Ecotourism, 4*(3), 178–195. https://doi.org/10.1080/14724040608668442

Blumberg, B., Peiro, J., & Roe, R. (2015). Trust and social capital: Challenges for studying their dynamic relationship. In F. Lyon, G. Mollering, & M. Saunders (Eds.), *Handbook of research methods on trust* (pp. 86–96). Edward Elgar Publishing.

Boksberger, P. E., & Craig-Smith, S. J. (2006). Customer value amongst tourists: A conceptual framework and a risk-adjusted model. *Tourism Review, 61*(1), 6–12. https://doi.org/10.1108/eb058465

Brown, J. A., Buchholtz, A. K., & Dunn, P. (2016). Moral salience and the role of goodwill in firm-stakeholder trust repair. *Business Ethics Quarterly, 26*(2), 181–199. https://doi.org/10.1017/beq.2016.27

Bundy, J., Pfarrer, M. D., Short, C. E., & Coombs, W. T. (2017). Crises and crisis management: Integration, interpretation, and research development. *Journal of Management, 43*(6), 1661–1692. https://doi.org/10.1177/0149206316680030

Butler, J. (1991). Toward understanding and measuring conditions of trust: Evolution of a condition of trust inventory. *Journal of Management, 17*, 643–663, https://doi.org/10.1177/014920639101700307

Çakar, K. (2020). Tourophobia: Fear of travel resulting from man-made or natural disasters. *Tourism Review, 76*(1), 103–124. https://doi.org/10.1108/TR-06-2019-0231

Cavlek, N. (2002). Tour operators and destination safety. *Annals of Tourism Research, 29*(2), 478–496. https://doi.org/10.1016/S0160-7383(01)00067-6

Chang, K.-C. (2014). Examining the effect of tour guide performance, tourist trust, tourist satisfaction, and flow experience on tourists' shopping behavior. *Asia Pacific Journal of Tourism Research, 19*(2), 219–247. https://doi.org/10.1080/10941665.2012.739189

Cheng, J. C., Chen, C. Y., Yen, C. H., & Teng, H. Y. (2017). Building customer satisfaction with tour leaders: The roles of customer trust, justice perception, and cooperation in group package tours. *Asia Pacific Journal of Tourism Research, 22*(4), 395–407. https://doi.org/10.1080/10941665.2016.1271816

Choi, M., Law, R., & Heo, C. Y. (2018). An investigation of the perceived value of shopping tourism. *Journal of Travel Research, 57*(7), 962–980. https://doi.org/10.1177/0047287517726170

Colquitt, J. A., LePine, J. A., Zapata, C. P., & Wild, R. E. (2011). Trust in typical and high-reliability contexts: Building and reacting to trust among firefighters. *Academy of Management Journal, 54*(5), 999–1015. https://doi.org/10.5465/amj.2006.0241

Connelly, B. L., Crook, T. R., Combs, J. G., Ketchen, D. J., & Aguinis, H. (2018). Competence- and integrity-based trust in interorganizational relationships: Which matters more? *Journal of Management, 44*(3), 919–945. https://doi.org/10.1177/0149206315596813

Czernek, K., & Czakon, W. (2016). Trust-building processes in tourist coopetition: The case of a Polish region. *Tourism Management, 52*, 380–394. https://doi.org/10.1016/j.tourman.2015.07.009

Han, W., Liu, W., Xie, J., & Zhang, S. (2022). Social support to mitigate perceived risk: Moderating effect of trust. *Current Issues in Tourism*, 1–16. https://doi.org/10.1080/13683500.2022.2070457

Hourigan, N. (2015). *Rule-breakers-why 'being there 'trumps' being fair' in Ireland: Uncovering Ireland's national psyche.* Gill & Macmillan Ltd.

Huang, S., Hsu, C. H. C., & Chan, A. (2010). Tour guide performance and tourist satisfaction: A Study of the package tours in Shanghai. *Journal of Hospitality & Tourism Research, 34*(1), 3–33. https://doi.org/10.1177/1096348009349815

Jackson, M. (2019). Utilizing attribution theory to develop new insights into tourism experiences. *Journal of Hospitality and Tourism Management, 38,* 176–183. https://doi.org/10.1016/j.jhtm.2018.04.007

Johnson, D., & Grayson, K. (2005). Cognitive and affective trust in service relationships. *Journal of Business Research, 58*(4), 500–507. https://doi.org/10.1016/S0148-2963(03)00140-1

Kapuściński, G., & Richards, B. (2016). News framing effects on destination risk perception. *Tourism Management, 57,* 234–244. https://doi.org/10.1016/j.tourman.2016.06.017

Kramer, R. (2009). Rethinking trust. *Harvard Business Review, 87*(6), 68–77.

Kumar, A., Capraro, V., & Perc, M. (2020). The evolution of trust and trustworthiness. *Journal of the Royal Society Interface, 17*(169), 20200491. https://doi.org/10.1098/rsif.2020.0491

Le Roux, T., & Van Niekerk, D. (2019). Challenges in stakeholders self-organising to enhance disaster communication. *Corporate Communications: An International Journal, 25*(1), 128–142. https://doi.org/10.1108/CCIJ-07-2019-0078

Legood, A., van der Werff, L., Lee, A., den Hartog, D., & van Knippenberg, D. (2022). A critical review of the conceptualization, operationalization, and empirical literature on cognition-based and affect-based trust. *Journal of Management Studies.* https://doi.org/10.1111/joms.12811

Mayer, R. C., Davis, J. H., & Schoorman, F. D. (1995). An integrative model of organizational trust. *Academy of Management Review, 20*(3), 709–734. https://doi.org/10.5465/amr.1995.9508080335

McAllister, D. (1995). Affect and cognition trust as foundations for interpersonal cooperation in organisations. *Academy of Management Journal, 38,* 24–59. https://doi.org/10.2307/256727

McTiernan, C., Musgrave, J., & Cooper, C. (2023). Conceptualising trust as a mediator of pro-environmental tacit knowledge transfer in small and medium sized tourism enterprises. *Journal of Sustainable Tourism, 31*(4), 1014–1031. https://doi.org/10.1080/09669582.2021.1942479

Meyerson, D., Weick, K., & Roferick, M. (2006). Swift trust in temporary groups. In R. Kramer (Ed.), *Organisational Trust.* Oxford University Press.

Morakabati, Y., Page, S. J., & Fletcher, J. (2017). Emergency management and tourism stakeholder responses to crises: A global survey. *Journal of Travel Research, 56*(3), 299–316. https://doi.org/10.1177/0047287516641516

Nooteboom, B. (2022). Trust. *Handbook on Theories of Governance* (pp. 205–214). Edward Elgar Publishing.

Nunkoo, R., & Gursoy, D. (2016). Rethinking the role of power and trust in tourism planning. *Journal of Hospitality Marketing & Management, 25*(4), 512–522. https://doi.org/10.1080/19368623.2015.1019170

Nunkoo, R., Ramkissoon, H., & Gursoy, D. (2012). Public trust in tourism institutions. *Annals of Tourism Research, 39*(3), 1538–1564. https://doi.org/10.1016/j.annals.2012.04.004

O'Malley, L., Harris, L. C., & Story, V. (2022). Managing tourist risk, grief and distrust post COVID-19. *Tourism and Hospitality Research, 23*(2), 170–183. https://doi.org/10.1177/14673584221089730

Okuyama, T. (2018). Analysis of optimal timing of tourism demand recovery policies from natural disaster using the contingent behavior method. *Tourism Management, 64*, 37–54. https://doi.org/10.1016/j.tourman.2017.07.019

Pagliara, F., Aria, M., Russo, L., Della Corte, V., & Nunkoo, R. (2021). Validating a theoretical model of citizens' trust in tourism development. *Socio-Economic Planning Sciences, 73*, 100922. https://doi.org/10.1016/j.seps.2020.100922

Park, S. (2020). Multifaceted trust in tourism service robots. *Annals of Tourism Research, 81*, 102888. https://doi.org/10.1016/j.annals.2020.102888

Patwardhan, V., Ribeiro, M. A., Payini, V., Woosnam, K. M., Mallya, J., & Gopalakrishnan, P. (2020). Visitors' place attachment and destination loyalty: Examining the roles of emotional solidarity and perceived safety. *Journal of Travel Research, 59*(1), 3–21. https://doi.org/10.1177/0047287518824157

Peccei, R., & Rosenthal, P. (2000). Front-line responses to customer orientation programmes: A theoretical and empirical analysis. *The International Journal of Human Resource Management, 11*(3), 562–590. https://doi.org/10.1080/095851900339765

Robert, L. P., Denis, A. R., & Hung, Y.-T. C. (2009). Individual swift trust and knowledge-based trust in face-to-face and virtual team members. *Journal of Management Information Systems, 26*(2), 241–279. https://doi.org/10.2753/MIS0742-1222260210

Rousseau, D., Sitkin, S., Burt, R., & Camerer, C. (1998). Not so different after all: A cross discipline view of trust. *Academy of Management Review, 23*(3), 393–404. https://doi.org/10.5465/amr.1998.926617

Sankowska, A. (2013). Relationships between organizational trust, knowledge transfer, knowledge creation, and firm's innovativeness. *The Learning Organization, 20*(1), 85–100. https://doi.org/10.1108/09696471311288546

Sarfraz, M., Raza, M., Khalid, R., Ivascu, L., Albasher, G., & Ozturk, I. (2022). Coronavirus disease 2019 safety measures for sustainable tourism: The mediating effect of tourist trust. *Frontiers in Psychology, 13*. https://www.frontiersin.org/article/10.3389/fpsyg.2022.784773

Seabra, C., Dolnicar, S., Abrantes, J. L., & Kastenholz, E. (2013). Heterogeneity in risk and safety perceptions of international tourists. *Tourism Management, 36*, 502–510. https://doi.org/10.1016/j.tourman.2012.09.008

Sparks, B. A., & Browning, V. (2011). The impact of online reviews on hotel booking intentions and perception of trust. *Tourism Management, 32*(6), 1310–1323. https://doi.org/10.1016/j.tourman.2010.12.011

Tomlinson, E. C., Schnackenberg, A. K., Dawley, D., & Ash, S. R. (2020). Revisiting the trustworthiness–trust relationship: Exploring the differential predictors of cognition- and affect-based trust. *Journal of Organizational Behavior, 41*(6), 535–550. https://doi.org/10.1002/job.2448

Ţuclea, C.-E., Vrânceanu, D.-M., & Năstase, C.-E. (2020). The role of social media in health safety evaluation of a tourism destination throughout the travel planning process. *Sustainability, 12*(16), Article 16. https://doi.org/10.3390/su12166661

van der Werff, L., & Buckley, F. (2017). Getting to know you: A longitudinal examination of trust cues and trust development during socialization. *Journal of Management, 43*(3), 742–770. https://doi.org/10.1177/0149206314543475

Weiler, B., & Black, R. (2015). The changing face of the tour guide: One-way communicator to choreographer to co-creator of the tourist experience. *Tourism Recreation Research, 40*(3), 364–378. https://doi.org/10.1080/02508281.2015.1083742

Williams, A. M., & Baláž, V. (2020). Tourism and trust: Theoretical reflections. *Journal of Travel Research*, 1–16. https://doi.org/10.1177/0047287520961177

Williams, A. M., Chen, J. L., Li, G., & Baláž, V. (2022). Risk, uncertainty and ambiguity amid Covid-19: A multi-national analysis of international travel intentions. *Annals of Tourism Research, 92*, 103346. https://doi.org/10.1016/j.annals.2021.103346

Xie, C., Zhang, J., & Morrison, A. M. (2021). Developing a scale to measure tourist perceived safety. *Journal of Travel Research, 60*(6), 1232–1251. https://doi.org/10.1177/0047287520946103

Zhong, L., Sun, S., Law, R., & Li, X. (2021). Tourism crisis management: Evidence from COVID-19. *Current Issues in Tourism, 24*(19), 2671–2682. https://doi.org/10.1080/13683500.2021.1901866

5 The evolving use of Generative AI in travel decision intelligence through periods of disruption

Kathryn Hayat

5.1 The evolution of traveller behaviour and decision making

The concept of travel decision making (often interchangeably used with the terms 'holiday' or 'tourist') has been on research agendas since the 1970s. The timing coincided with the birth of modern, consumer-led travel in the Western world, and the research was used as a tool to explore and understand the phenomena of newly emerging mass, organised travel and the patterns of this within Western markets. The research, based on established consumer decision making and purchase behaviour models within American marketing circles, resulted in early travel-focused models that were centred on the simplistic understandings of a typical person's behaviour (often a middle-class, white Western male/nuclear family). One noted Model of the Travel Decision Process published by Schmoll (1977) proposed four unconnected factors that played a role in the decision of the consumer. The four factors focused on travel stimuli, personal determinates, external variables and characteristics of a destination, simplified as push and pull factors. The model, which set out no order of significance for the factors, offered an early insight into the wide range of influences on a travel-related decision. In the following decade, Mathieson and Wall (1983) proposed the Travel Buying Behaviour Model, using some of the factors established in Schmoll's (1977) model to propose clear, linear stages that roughly equated to pre (the decision stages from idea/inspiration to booking), during (whilst travelling and in situ) and post (reflecting and memories of) travel. This model clearly identifies a shift in just five years in the understanding of traveller behaviour from unconnected to interconnected influential factors. In the decades since, numerous authors have proposed many models that reflect changes to Schmoll's (1977) four unconnected factors, recognising a range of interdependent, transient and salient factors, as well as nuances based on socio-cultural differences and a wide range of disrupters.

To reflect an increasingly international tourism market established through emerging markets and economic blocs, and include all types of increasingly unique travel and travellers, there are now a range of continually published travel behaviour and decision-making models that reflect unique scenarios across international markets and consumer groups. For example, the influence

DOI: 10.4324/9781032720555-6

of a family unit presented by Koc (2004), identified the influential role women in Turkey have in purchasing tasks and suggested the recommendation for elevated relevance of this to tourism marketers. The model identifies the wider benefits of understanding a specific behaviour for a range of stakeholders. Almost a decade later Bronner and de Hoog (2011) added a further dynamic to travel buying behaviour by identifying the influence of eWOM (electronic Word of Mouth) on other travellers. With a different focus, this research reflects changes through the 2000s in internet use and behaviour through Web 2.0 technology and user-generated content platforms. Findings suggested that the reviews posted to independent review websites such as Trip Adviser were significant for tourism marketers, but also disruptors to the market, leading to more complex and elongated decisions for travellers overall. Of course, reflecting every single individual difference for every traveller would create an unnecessary number of unique models and propositions, defeating the original purpose of the models to simplify knowledge. However, the greater understanding all stakeholders have of the dynamic, changing parameters of travel buyer behaviour and decision making can be seen to only support the continued effectiveness of industry stakeholders.

5.2 The influence of disruption on traveller behaviour and decision making

Changes in traveller behaviour and decision making due to evolutions in technology or societal norms are well reflected in research. Furthermore, the tourism industry and associated sectors, and stakeholders within, will already be familiar with the patterns and nuances of different market segments through general market knowledge and business intelligence. Website and app design, marketing tactics, branding and customer service standards will all, to an extent, reflect these acknowledged stages in the behaviour of a tourist to compete and be attractive to users. However, what is less researched is the influence of macro-environmental changes and the potential resulting disruption from these in the shorter and longer term and any impact of these changes on traveller behaviour and decision making. The only exception tends to be research into unexpected transport disruption such as public transport across Europe (Eltved et al., 2021; Gasparinatou et al., 2023), where there are arguably more localised political and economic ramifications to such disruption in the immediate term. Turbulence globally during the last two decades, often perceived as heightened and unexpected due to internationally impactful terrorist activity, global health endemics (escalated by the Covid-19 Pandemic) and political instability, has changed the dynamics of accepted transient and salient factors that influence the decision making and behaviour of tourists. Furthermore, the unpredictability of many of the turbulent macro events can make the influencing factors arise quickly, with little time for travellers or stakeholders to understand and react. Equally, disruptive events can disappear as quickly as they emerge, as the turbulence becomes managed better by stakeholders and accepted more by travellers (Rhama, 2022) and to an extent normalised by the travel industry and travellers alike. McKercher (2021), for example, suggests that many destinations experiencing disruption not only return to travel quite quickly but also experience growth, often as a surprise to many, with the 9/11 Twin Towers terrorist attack in New York, USA, often cited as a turning point in recovery and growth speed post-disruption. It is acknowledged by Karl et al. (2021) that affective decision making – the emotions and mental predictions humans use to aid decision making – can be influenced negatively because of anxiety. For example, travellers may perceive higher risk than usual and avoid any risk regardless of the potential gains. Therefore, during times of disruption where anxiety is likely to be greater than usual, affective influences can result in negative impacts on decision making, with the extreme negative impact being the decision not to travel at all. However, traveller behaviour has become more controlled by the traveller in recent years through changes to the market as a result of the autonomy created and facilitated through technological evolution. For example, the concept of the Sharing Economy made famous by the now household brands of Airbnb and Uber uses a Peer-to-Peer (P2P) business and economic model that, through enabling technology, is centred around traveller-to-traveller interactions. These interactions can be naturally more agile (such as quicker cancellations, refunds and personal communication between users often using messaging platforms similar to those used in family and friend interactions). This often means that P2P businesses are more responsive to disruption (for example, a town experiences a flood and a hotel can send an instant message to the tourist with implications, alternatives or resolutions). The simplicity of a message can lead to a reduction in anxiety and reduce or remove stressors for travellers. Social media platforms and the vast content libraries within these and other user-generated content media are now easily available to global markets. A search engine result or a vertical search within a platform (for example, searching with Instagram or TikTok for a phrase or hashtag) provides travellers with access to a wide multimedia of information (images, videos, live-streamed content, etc.) from which they can make decisions and adapt behaviour.

Past research (Buhalis et al., 2019) has illustrated the influence that such technological change and development can have on traveller behaviour. However, this is often limited in scope and tends to reflect the notion that traveller decision making is a planned behaviour based on rational choices (Stylos, 2020) from several set alternatives, mirroring the simplistic models from the 1970s (such as Schmoll, 1977; Mathieson & Wall, 1983), already discussed. However, the increase in the pace of technological change in recent years has seen the explosion of enabling technology that Buhalis et al. (2019, p. 487) referred to as fifth-generation mobile network (5G); artificial intelligence (AI); radio frequency identification (RFID); mobile devices, smartphones and wearables; applications or apps (along with APIs [Advance Passenger Information]); cryptocurrency and blockchain, and is reflective of changes in consumer, and therefore, travel behaviour. These enabling technologies are all deemed to create added value in service marketing and management but are used in tandem as both consumer devices and technologies, as well as having a business function. Whilst the use of much of these enabling technologies can be considered widespread and without global barriers, it is important to acknowledge Smallman and Ryan's (2020) distinction in the difference in digital skills between younger (Generation Y) and older millennials (Generation Z) and its relevance to differences in travel behaviour and adoption of the enabling technologies mentioned above – a nuance that can be explored through a personalised understanding of travel behaviour and decision making.

Recognising AI as one of the five key enabling technologies in 2019, Buhalis recognised the fast pace at which the technology was developing. In the years since, the technology has experienced a significant change in usage propelled by its use in image recognition (such as Snapchat filters and viral photo-based applications), decision-making (such as weather predictions and live activity/busy times) and automated tasks (smart home applications such as light bulbs and white goods). As a result, social adoption of AI has happened at a very rapid rate in a period of what Samala et al. (2022) refer to as technology-driven times. Perhaps as an evolution from the internationally adopted and well-established use of social media (such as Facebook or Weibo) and P2P business models (such as Airbnb or Uber), AI creates both a range of opportunities and challenges for stakeholders as once again traveller decision making is influenced by additional transient and salient factors. Across the hospitality and tourism industries, service operations have quickly established uses of AI in a range of aspects of the tourism experience (generally focused on pre-, during and post-travel). For example, chatbots are commonly integrated within a website or app at the pre-booking and travel stages and have become an established form of communication for many travellers, particularly useful in addressing problems or complaints (examples include Skyscanner, Accor Hotels and KLM). Research by Pillai and Sivathanu (2020) concluded that the perception of chatbots was positive, the anxiety of users was minimal and, in general, chatbots have become a quickly adopted technology. Robot concierge and self-service kiosks at a hotel or airport during travel are commonly used and accepted alternatives to human service, with the first airlines offering mainstream self-service in the early 2000s (Airport Business, 2012). Post-travel, AI can create memories from traveller data in bespoke photo book stories or movies often directly integrated into a social media platform. However, the accuracy and authenticity of the memories can be questioned, given the increasingly sophisticated filtering qualities of AI tools, leading to a range of ethical considerations.

The adoption of Large Language Models (LLMs), systems powered by AI through vast amounts of data, offer personalised suggestions for users. Mainstream examples, such as Google's autocomplete predictions in the search box, are something millions of travellers will already be familiar with, often ignorant to the fact that they are already accustomed to using AI so much in daily life. Rapid AI developments in LLM mean that tools such as Open AI's ChatGPT and Google's Gemini have enhanced this kind of personalisation and can be seen as somewhat revolutionising human resembled communication. Kim et al. (2023) refer to this as a paradigm shift in how information is both acquired and used. This has occurred in synergy with generational change in recent years, with Generation Z (or the Alpha Generation) increasingly symbiotic in the use of technology and digital applications as part of tourism practice and behaviour. Immersive technology such as Virtual Reality (VR), Augmented Reality (AR) and live experiences-often commonly referred to overall as the metaverse has created endless ways to experience tourism in a way that is inseparable from technology. Perhaps because of this established immersive technology, AI-enabled technology has quickly been accepted and adopted in a comparably short period of time to previous technologies to assist and communicate with tourists (Samala et al., 2022) with features such as responsiveness, flexibility and availability of all now commonly expected as standard from technological solutions.

5.3 Significance of Generative AI to recovery and resilience

Gault et al. (2019) refer to AI as an evolution from social media networks as the established technology of choice to use during times of disruption. What AI offers as an extension to social media is a highly tailored technological tool that becomes a way of offering more personalised alerts to travel during these periods of disruption. Publicly available big data, such as social media posts during times of disruption in the past combined with location data collected via GPS to provide busy times, travel delays and incidents, is already used to achieve smart mobility. Understanding where AI and future technological advancements that aid decision making sit within the tourism ecosystem is paramount to support understanding of how recovery and resilience through times of disruption can be aided and allow stakeholders to understand the influence and potential ramifications on travel behaviour and decision making. What AI can contribute through machine learning is a combination of insights into past experiences, plus live reactions and the ability to support continuous decision making for the traveller during periods of disruption. For example, Generative AI provides travellers with easily accessed and user-friendly tools to support reactions to changes and unexpected issues proactively. AI tools such as ChatGPT and Gemini are increasingly integrated into day-to-day activities with more sophisticated outputs and continually more accurate information. Transient and salient factors emerging because of disruption can be managed better by the traveller with the additional support provided via Generative AI, contributing to making risk-related decisions less stressful (Zhu et al., 2021). Whilst the value of AI tools is still being questioned in terms of issues such as accuracy, accountability and data bias, there is undoubtedly a place for AI tools to enhance the experience of travellers with 'engaging, personalized and cost-effective services' (Kim et al., 2023). The ease of accessing information from AI can be demonstrated through searches on the ChatGPT platform. Figures 5.1–5.3 provide a sample of real responses from ChatGPT to fictional scenarios and commands relating to travel disruption pre-, during and after

I booked a ryanair flight with Love Holidays but I cannot log in to add extra luggage. How can I do this?

To add extra luggage to your Ryanair flight booked through Love Holidays, follow these steps:

1. **Retrieve Your Ryanair Booking Reference:**

   - Check your booking confirmation email from Love Holidays. They should have provided a Ryanair booking reference, usually a 6-character alphanumeric code.

2. **Visit Ryanair's Website:**

   - Go to the Ryanair website: Ryanair.com.

3. **Manage My Booking:**

   - Click on the "My Bookings" tab at the top of the homepage.
   - Select "Manage my booking."

4. **Log In or Retrieve Booking:**

   - If you have a Ryanair account, log in using your email and password.
   - If you do not have an account or prefer not to log in, select "Don't have an account? Find my booking" or a similar option.

5. **Enter Your Details:**

   - Enter your Ryanair booking reference and the email address used for the booking (this might be the email used by Love Holidays to book the flight).

6. **Access Your Booking:**

   ↓

   - Once you access your booking, you should see an option to add or modify baggage. Follow

ChatGPT can make mistakes. Check important info.

Figure 5.1 ChatGPT response 1.

travel. Whilst the information in each response is based on widely available information online, the information is presented as a personal response to a question, importantly a single user-friendly response drawing information from various sources and provided in a more humanistic way almost instantly. These are features that could all be used by travellers and stakeholders to contribute to reducing travel stress and anxiety during periods of disruption.

By reducing travel anxiety during periods of disruption, the decision making and behaviour of travellers can be stabilised as AI facilitates continuous travel decision making. Travellers can be more reactive, and rather than

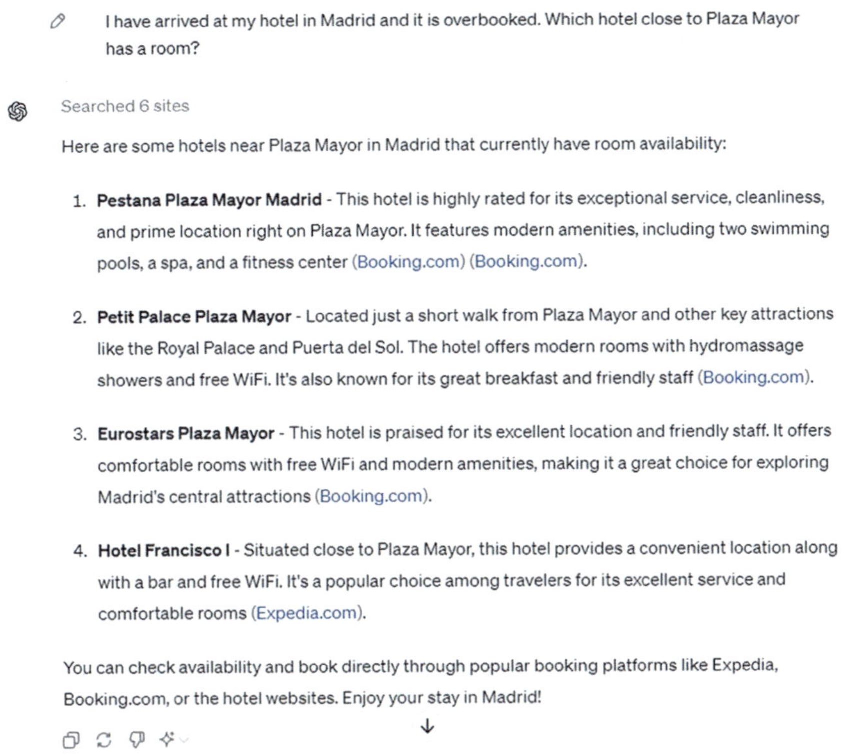

I have arrived at my hotel in Madrid and it is overbooked. Which hotel close to Plaza Mayor has a room?

Searched 6 sites

Here are some hotels near Plaza Mayor in Madrid that currently have room availability:

1. **Pestana Plaza Mayor Madrid** - This hotel is highly rated for its exceptional service, cleanliness, and prime location right on Plaza Mayor. It features modern amenities, including two swimming pools, a spa, and a fitness center (Booking.com) (Booking.com).

2. **Petit Palace Plaza Mayor** - Located just a short walk from Plaza Mayor and other key attractions like the Royal Palace and Puerta del Sol. The hotel offers modern rooms with hydromassage showers and free WiFi. It's also known for its great breakfast and friendly staff (Booking.com).

3. **Eurostars Plaza Mayor** - This hotel is praised for its excellent location and friendly staff. It offers comfortable rooms with free WiFi and modern amenities, making it a great choice for exploring Madrid's central attractions (Booking.com).

4. **Hotel Francisco I** - Situated close to Plaza Mayor, this hotel provides a convenient location along with a bar and free WiFi. It's a popular choice among travelers for its excellent service and comfortable rooms (Expedia.com).

You can check availability and book directly through popular booking platforms like Expedia, Booking.com, or the hotel websites. Enjoy your stay in Madrid!

Figure 5.2 ChatGPT response 2.

significantly changing decisions or behaviour (such as cancelling travel plans or boycotting a destination), they can continue in a modified way, thus becoming more resilient overall. This can be of benefit to travel stakeholders as the focus at times of disruption shifts from providing a reactive response to a disruption to focusing on efforts relating to recovery from and resilience to disruption. There are already some early interesting established examples of how AI can support recovery and resilience across the tourism and hospitality industries. Modgil et al. (2022) identified how the speed of information flow through AI technology can lead to a greater recovery capacity as the resulting supply chain has increased resilience. Furthermore, Samala et al. (2022) suggest that longer-term growth can be gained through AI tools as the tool becomes a competitive advantage or unique selling point in a market. Specifically, within aviation, IATA (2023) focuses on the scope of Generative AI for disruption management as revolutionary, with the potential to influence travel plans, disruption management and support, all personalised at the point of delivery.

Through times of disruption, the usefulness can be even greater when issues are often unexpected and require quick actions from stakeholders to maintain customer satisfaction and expedite overall recovery and resilience. Increasingly,

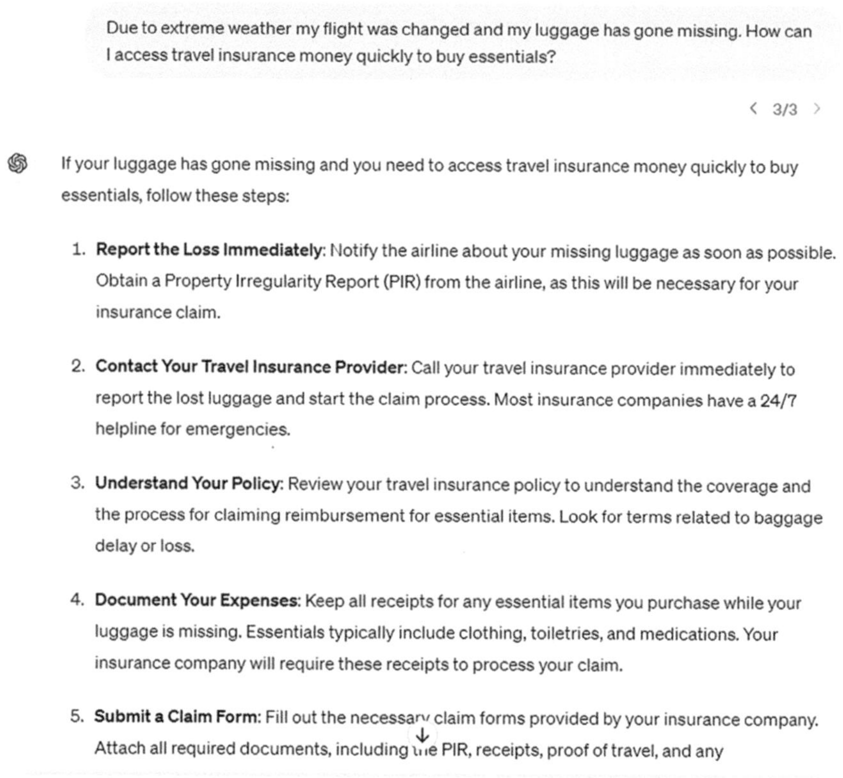

Due to extreme weather my flight was changed and my luggage has gone missing. How can I access travel insurance money quickly to buy essentials?

‹ 3/3 ›

If your luggage has gone missing and you need to access travel insurance money quickly to buy essentials, follow these steps:

1. **Report the Loss Immediately**: Notify the airline about your missing luggage as soon as possible. Obtain a Property Irregularity Report (PIR) from the airline, as this will be necessary for your insurance claim.

2. **Contact Your Travel Insurance Provider**: Call your travel insurance provider immediately to report the lost luggage and start the claim process. Most insurance companies have a 24/7 helpline for emergencies.

3. **Understand Your Policy**: Review your travel insurance policy to understand the coverage and the process for claiming reimbursement for essential items. Look for terms related to baggage delay or loss.

4. **Document Your Expenses**: Keep all receipts for any essential items you purchase while your luggage is missing. Essentials typically include clothing, toiletries, and medications. Your insurance company will require these receipts to process your claim.

5. **Submit a Claim Form**: Fill out the necessary claim forms provided by your insurance company. Attach all required documents, including the PIR, receipts, proof of travel, and any

Figure 5.3 ChatGPT response 3.

the use of AI acting as a pseudo trip assistant through all stages of buyer behaviour can provide personalised information, something stakeholders often find challenging to offer through human-to-human interaction. This becomes poignant given the acknowledged widespread staffing shortages across the sectors post-pandemic. It is, however, important to recognise that Generative AI is still evolving, and the LLMs that underpin the increasingly household AI names are constantly learning and evolving from human input. This means that AI can and does produce flawed responses. Sometimes these are obvious but not always, which raise questions of accuracy. Ultimately, the content generated via AI is not a human response but human-like. To continue building user trust and adoption, ensuring accurate and dynamic data, that IATA (2023) states is crucial to adoption and success, will be key to overcoming any challenges.

It can be suggested that the more AI tools like ChatGPT are used, the greater the potential of perceived usefulness and ease of use. AI should be viewed as both a tool to support the traveller in situ and as a tool to aid tourism recovery through multiple ways, such as reducing longer-lasting disruptions, alleviating anxiety and stress for travellers (ultimately becoming more

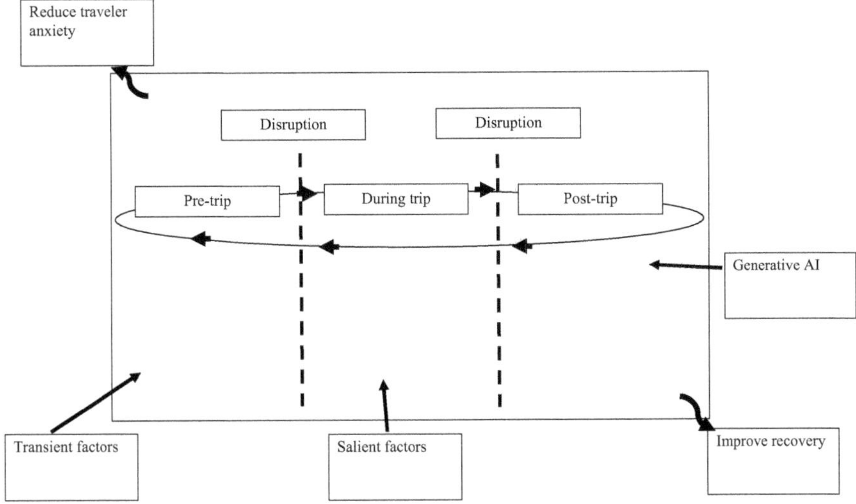

Figure 5.4 AI's future place in the tourism ecosystem.

self-sufficient) and diverting tourism recovery efforts away from customer support as travellers become more independent. Disruption becomes more normalised/expected in travel, but with reduced attached risks. This can be seen in Figure 5.4, which identifies the place that AI is expected to take in the near future within the tourism ecosystem. It is suggested that at all stages of travel buyer behaviour and decision making, AI plays an important and influential role in mitigating transient and salient factors. Not only can this reduce anxiety and stress for the traveller during times of disruption, but it can also build resilience to disruption at both the traveller and stakeholder levels, thus resulting in expedited recovery from disruption all around.

5.4 Conclusion

The evolution of the phenomena of modern tourism in the last century has proven that new and emerging markets both enjoy the benefits of inbound tourism and flourishing tourism sectors, alongside the appetite for tourism activity from citizens. As the peripheries of tourism have spread out from the Western world to now established nation-states and economic blocs globally, so has the diversity in both supply and demand for travel, hospitality and wider tourism activity. This expansion in the accessibility of travel and tourism on an international scale, undoubtedly, leads to increasingly more complex decision making and travel behaviour, offering a stark contrast in comparison to the birth of modern-day travel in the 1970s and the understanding of such behaviour and decision making. As a global society now well-established in its use of fifth-generation enabled technologies, such as the internet, social media platforms and AI, the scope for the use of Generative AI tools in serving the travel community is endless. The machine learning subset of AI and the vast human usage of AI provide the scope for more human-like AI functionality, which societies are increasingly becoming used to in day-to-day technology. As the number of global travellers continues to grow, particularly from emerging markets, the increasingly efficient and effective use of AI becomes increasingly attractive to both tourists and suppliers alike. With more and more people travelling in more dynamic and personalised ways, less standardised than at the time of publication of many established traveller decision making and buyer behaviour models, AI offers many prospective benefits. Using this technology as an aid during times of disruption to support continuous decision making can help reduce the anxieties and stresses of travellers. This could be used through periods of discrete disruption, such as air space closure, natural disasters or health pandemics, when tourists and suppliers globally face challenges. It could also benefit day-to-day travel too, ensuring that the pleasures of travel that make the industry so attractive remain at the forefront of travel behaviours and decision making on a more modest scale. The wider ramifications for the recovery and resilience of travel stakeholders are yet to be seen as the technology evolves and usage increases, but it could support the speed and impact of business recovery, the reopening of a tourist space and the resilience of a tourism market. The potential to view Generative AI as an important influencing factor in the travel ecosystem is great and one to be continually explored as technological development and widespread use continue.

References

Airport Business (2012, March). *The evolution of the kiosk*, http://www.airport-business. com/2012/06/the-evolution-of-the-kiosk/

Bronner, F., & de Hoog, R. (2011). Vacationers and eWOM: Who posts, and why, where, and what? *Journal of Travel Research*, 50(1), 15–26. https://doi.org/10. 1177/0047287509355324

Buhalis, D., Harwood, T., Bogicevic, V., Viglia, G., Beldona, S., & Hofacker, C., (2019). Technological disruptions in services: Lessons from tourism and hospitality. *Journal of Service Management*, 30(4), 484–506. https://doi.org/10.1108/JOSM-12-2018-0398

Eltved, M., Breyer, N., Ingvardson, J. B., & Nielsen, O. A. (2021). Impacts of long-term service disruptions on passenger travel behaviour: A smart card analysis from the Greater Copenhagen area. *Transportation Research Part C: Emerging Technologies*, 131, 1–16. https://doi.org/10.1016/j.trc.2021.103198

Gasparinatou, C., Mantouka, E., Vlahogianni, E. I., Masegosa, A. D., & Serrano, L. (2023). Travel behavior shifts under extreme system-level disruptions. *Transportation Research Procedia*, 72, 1420–1426. https://doi.org/10.1016/j.trpro.2023.11.606.

Gault, P. Cottrill, C. D. Corsar, D. Edwards, P. Nelson, J. D. Markovic, M., Mehdi, M., & Sripada, S. (2019). TravelBot: Utilising social media dialogue to provide journey disruption alerts. *Transportation Research Interdisciplinary Perspectives*, 3, 1–9. https://doi.org/10.1016/j.trip.2019.100062

IATA (2023, December). *Generative AI and aviation: Finding crossroads for future implementation*, https://www.iata.org/globalassets/iata/programs/innovation-hub/ generative-ai-report.pdf

Karl, M. Kock, F. Ritchie, B. W., & Gauss, J. (2021). Affective forecasting and travel decision-making: An investigation in times of a pandemic. *Annals of Tourism Research*, 87, 103–139 https://doi.org/10.1016/j.annals.2021.103139

Kim, J. H., Kim, J., Kim, S., & Hailu, T. B. (2023). Effects of AI ChatGPT on travelers' travel decision-making. *Tourism Review*, Vol. ahead-of-print No. ahead-of-print. https://doi.org/10.1108/TR-07-2023-0489

Koc, E. (2004). The role of family members in the family holiday purchase decision making process. *International Journal of Hospitality and Tourism Administration*, 5(2), 85–102. https://doi.org/10.1300/J149v05n02_05

Mathieson, A., & Wall, G. (1983). Tourism: Economic, physical, and social impacts. Longman Group Limited. *Journal of Travel Research*, 22(1), 51–51. https://doi.org/10.1177/004728758302200113 1

McKercher, B. (2021). Can pent-up demand save international tourism? *Annals of Tourism Research Empirical Insights*, 2(2), 1–4. https://doi.org/10.1016/j.annale.2021.100020

Modgil, S., Gupta, S., Stekelorum, R., & Laguir, I. (2022). AI technologies and their impact on supply chain resilience during COVID-19. *International Journal of Physical Distribution & Logistics Management*, 52(2), 130–149. https://doi.org/10.1108/IJPDLM-12-2020-0434

Pillai, R., & Sivathanu, B. (2020), Adoption of AI-based chatbots for hospitality and tourism. *International Journal of Contemporary Hospitality Management*, 32(10) 3199–3226. https://doi.org/10.1108/IJCHM-04-2020-0259

Rhama, B. (2022). Local communities' and tourists' adaptation to pandemic-induced social disruption: Comparing national parks and urban destinations. *International Journal of Disaster Risk Reduction*, 82, 1–18. https://doi.org/10.1016/j.ijdrr.2022.103380

Samala, N., Katkam, B. S., Bellamkonda, R. S., & Rodriguez, R. V. (2022). Impact of AI and robotics in the tourism sector: A critical insight. *Journal of Tourism Futures*, 8(1), 73–87. https://doi.org/10.1108/JTF-07-2019-0065

Schmoll, G. (1977). *Tourism Promotion* (1st ed.). Tourism International Press, London.

Smallman, P., & Ryan, C. (2020). Order out of chaos: Nudging tourists' behaviours at a time of disruption? *Journal of Tourism and Hospitality Management*, 8(4), 131–139. https://doi.org/10.17265/2328-2169/2020.04.001

Stylos, N. (2020). Technological evolution and tourist decision-making: A perspective article. *Tourism Review*, 75(1), 273–278. https://doi.org/10.1108/TR-05-2019-0167

Zhu, T. Haugen, S., & Liu, Y. (2021). Risk information in decision-making: Definitions, requirements and various functions. *Journal of Loss Prevention in the Process Industries*, 72, 1–13. https://doi.org/10.1016/j.jlp.2021.104572

6 How to Respond When the Sky Closes Due to a Technical Issue? Navigating Air Travel Shutdowns

Effective Response Strategies to Technical Disruptions

Adam Jones, Elyssa Fanning and Geri Silverstone

6.1 Introduction – Contextualising the Crisis: The August 2023 UK Airspace Shutdown

Safety should be at the heart of any aviation system (Murphy and Efthymiou, 2017). On 28th August 2023, due to a major failure of the flight planning system, the UK airspace was effectively shut down to maintain safe operations. The shutdown occurred on an August bank holiday, traditionally among the busiest flight days in the UK, and impacted one of the most complex and congested airspaces in the world (Foster et al., 2021). This technical failure impacted hundreds of thousands of passengers over a number of days and cost millions of pounds. For lessons to be learnt and resilience to be developed, there is a need to understand what occurred and propose recommendations that will support future effective response strategies to technical disruptions.

Under normal circumstances, the UK flight planning system processes approximately 800 flight plans per hour. As a result of the system failure, there was a need to revert to manually inputting flight plans, reducing the number processed to approximately 60 per hour. The impact was compounded by Heathrow and Gatwick, two of the busiest airports in Europe, both running at virtually total capacity. The system shutdown created a major incident across the aviation network that affected airports, airlines, tour operators and, most importantly, their passengers. Without the automated flight planning system, air traffic controllers could handle far fewer flights – just 15% of the normal flow. Aircraft already in flight were able to continue without being made to divert. Still, with the limited number of flight plans being processed, most departures were delayed or cancelled with planes kept on the ground.

Worldwide passenger numbers are expected to increase annually by 3.8%, resulting in an estimated additional 4 billion passenger journeys in 2043 compared to 2023 (IATA, 2024). This increase will provide additional demand on an already complex system that is operating at capacity. Through analysing the Civil Aviation Authority's Independent review of the flight planning system, written and oral evidence given to the Government's Transport Select Committee, along with media articles and social media posts, this chapter

DOI: 10.4324/9781032720555-7

explains the system's failure and analyses its socioeconomic and operational impacts. The chapter explores the sector's responses to the impacts of such a significant reduction in flight airspace and then details and evaluates the implications of such a technical failure. Through analysis using crisis management and communication theory, the chapter concludes by suggesting how future resilience may be built and provides strategic recommendations through the lessons learnt.

6.2 Mechanisms of Air Traffic System Failure and Resulting Disruptions

Airlines planning to operate flights through controlled airspace must file a flight plan containing information such as aircraft type, speed and routing. This information is required by various Air Navigation Service Providers (ANSPs) who provide Air Traffic Services (ATS) during the flight. NATS is the UK's leading ANSP provider of ATS.

The UK participates in the Integrated Initial Flight Plan Processing System, part of the Eurocontrol centralised Air Traffic Flow Management system. Once received from Eurocontrol, flight plans are initially processed by NATS to identify the portion of the route to be flown in UK airspace and to extract the data for presentation to controllers at their workstations via the UK National Airspace System (NAS). Processed flight data is presented to the system four hours before being required by the relevant air traffic controller or airspace sector. The presentation of flight data is provided continuously so that, at any given moment, NAS will contain flight data for the following four-hour period.

At 4:59 am on 28th August 2023, French Bee flight BF731 departed from Los Angeles (LAX), heading to Orly Paris (ORY). The flight path was scheduled to go over the western USA, eastern Canada, Northern Ireland and the UK, making landfall in Deauville, France and finally landing in Paris. The flight plan contained the navigational beacons, given as three-letter codes, some of which were duplicates. The code for Deauville, France (DVL), is the same as Devil's Lake in North Dakota, USA. The flight plan was shared by Eurocontrol at 8:32 am with air navigation services, including NATS in the UK.

On 28th August 2023 the duplicate code for Deauville, France and Devil's Lake in North Dakota, USA, caused a problem. The NATS system processed this flight plan data by indicating that the plane would be landing in North Dakota (USA) rather than Deauville (France), which meant the aircraft would need to leave UK airspace before it had even entered it. The system went into maintenance mode to prevent the transfer of what appeared to be corrupt flight data. The backup second system processed the codes for the flight plan in the same way and also went into maintenance mode. Both NATS systems had stopped automatically processing flight paths, which meant all the coding had to be undertaken manually, triggering devastating effects for many air travellers hoping to leave the UK or travel on the bank holiday Monday.

The recovery process exacerbated the incorrect flight data and the problems it caused, with several different NATS system engineers trying to resolve the issue over a prolonged period. Over four and a half hours after the initial failure, an engineer from the software supplier was called, who understood the error and rectified the problem.

6.3 Analysing the Socioeconomic and Operational Impacts of the Disruption

The total number of cancelled flights remains unclear, but it was considerable. A cancelled or significantly delayed flight can lead to aircraft and crew being displaced and unable to operate the anticipated return service. For safety reasons, the flight crew must adhere to the minimum regulatory requirements for rest. Such issues can disrupt an airline's operation and make quantifying the exact number of flights impacted difficult. However, it has been estimated that when cancellations on the day are combined with the impact of airlines reorganising themselves to recover their flight schedules, the total number of cancellations exceeds 2,000 flights (Transport Select Committee, 2023j). In addition, out of the 5,500 flights operating in UK airspace on 28th August, approximately 10% experienced delays due to the incident.

The incident disrupted the journeys of more than 700,000 passengers (CAA, 2024). The CAA estimates that over 300,000 passengers were impacted by flight cancellations, approximately 95,000 by long delays (over three hours) and at least a further 300,000 by shorter delays (CAA, 2024). The last flight to be affected was on 4th September, eight days after the initial incident (Calder, 2024).

When flights are severely delayed or cancelled, airlines are required to provide care and assistance to passengers. While passengers wait at the airport, this provision includes refreshments, and where necessary, airlines must also offer overnight accommodation. Airlines must also inform passengers of their legal rights and offer multiple and effective means of communication for passengers who need to get in touch. When a flight is cancelled rather than just delayed, airlines are required to offer passengers a choice of a full refund, paid within seven days, or the option of being rerouted at the earliest opportunity or at an agreed date via any carrier or alternative transport mode. Any later alternative flight could entail additional overnight accommodation, transport and food, which the airline is liable to reimburse.

The following personal travel stories bring to life the impacts of the NATS technical issue. Such narratives provide a micro review of the implications of such travel disruption on health, life events and daily routines:

- Serena Hamilton, from Northern Ireland, was unable to attend her Newcastle hospital check-up following a heart attack (Wilson, 2023).
- For Adam Ashall-Kelly and his fiancée, the disruption would affect not only them but also their 45 wedding guests. They were concerned about how the cancellation of their Manchester flight would impact their planned Italian

wedding, with their friends and family being required to arrange new flights from different airports across the UK (BBC, 2023).

- Rather than wait a week for a rescheduled flight offered by easyJet, Mark and Holly Baker had to consider what was best for their family. They took an overnight ferry with their children to Toulon, France, then three trains and finally another ferry to get to their home in Brighton to ensure their daughter started secondary school on time (Race, 2023).

6.4 Crisis Management Responses: Stakeholder Communication and Operational Adjustments

Airports and airlines were informed, alerted or made aware of the disruption to the air traffic control system through various sources and channels. This communication mix was a concern to those affected, especially when there were only sometimes official notifications from NATS. Tim Alderslade, CEO of Airlines UK, the trade body for UK-registered airlines, stated that there was 'some very poor communication from NATS. We did not hear from them formally until the following day…I heard about it from Sky News' (Transport Select Committee, 2023b, Q18). The CAA's (2024) incident timeline of the day identifies that the first airline made aware of the issues was Tui, informed at 10:04 am through their Hanover Group Operations Centre. Virgin Atlantic was the next airline through a message on the Eurocontrol portal at 11:00 am. At 11:07 am easyJet received a call from Eurocontrol informing them that flight movements in the UK would be limited. British Airways reported a similar story. Local NATS operations informed individual airports of the situation.

Official notification for most major airlines and airports was eventually received directly from NATS via an email at 11:30 am, almost three hours after the first sign of a systems issue (CAA, 2024). The email stated that the Air Traffic Incident Communication and Coordination Cell (ATICCC) had been activated. ATICCC allows the publication of necessary information relating to an event and NATS' response to it in the form of text updates and stakeholder teleconferences. This information provided the opportunity for airlines and airports to hear updates directly from NATS regarding the current situation and NATS' expectations for the immediate future. This official notification is designed to support airlines and airports in planning their own operations.

The challenge for airlines with such late notification was that it became more difficult to recover and accommodate the revised flying programme. This three-hour window from 8:00 am is seen as critical by airlines and their representatives. Tim, CEO of Airlines UK, commented, 'That really impacted our ability to contact customers and to put in place contingency arrangements around hotels and the like, as part of our obligation' (Transport Select Committee, 2023a, Q17). Similarly, Sophie Dekkers, Chief Financial Officer of easyJet, commented that the delay in notification increased the 'challenge of recovery and trying to predict and understand the flying programme and its impact was very significant' (Transport Select Committee, 2023e, Q22).

At 12:30 pm, Gatwick Airport requested airlines to cancel 80% of flights and close their check-in desks. Shortly after this, Heathrow also asked airlines to cancel all UK, Ireland and European flights until 6:00 pm. It was not until 3:00 pm, seven hours after the system failure that the ATICCC was deactivated, and NATS informed airlines that the system had returned to normal. All traffic restrictions were removed by 6:00 pm that evening; however, airports remained extremely busy on Tuesday, 29th August 2023, due to the impact of the many cancellations of the previous day (CAA, 2024).

To cater for the disruption of the most impacted destinations by the technical issue, airlines tried to provide extra seats by scheduling additional flights or upgrading flights to larger aircraft. However, such options were not feasible to cater for the full impact of the disruption. Easyjet reported over 110,000 passengers were affected, which would equate to a requirement of around 100 aircraft to recover them all (Transport Select Committee, 2023e, Q22). In the days following the incident, airlines and airports were focused on recovering their schedules, repositioning aircraft and ensuring that flight crews were back in the correct locations to operate the services.

Information dissemination proved difficult, with many passengers reporting they received more information through online and social media than from their airlines. On a typical day, airlines handle hundreds of flights, but with such extensive cancellations and delays, the process of managing refund and reroute requests was significantly slower, with telephone lines and email communications being quickly overwhelmed. Social media channels provided alternative options, and many airlines reported that their teams were working around the clock to respond to customer queries and concerns.

Airlines responded with their usual communication method: passengers who booked directly with the airlines received an email setting out in full their entitlements, their right to compensation and their right to travel on alternative airlines. Where possible, customers were also notified by text message and via push notification on the airlines' apps (Transport Select Committee, 2023e, Q22). Where customer details were not available, notification was through their booking agent, which caused a delay in people understanding what was happening.

To support those passengers who may not have access to or be confident using mobile phone apps, airlines set up dedicated contact centres, proactively calling those who had not been able to amend their booking and informing them of their options. The teams also made outbound calls, rather than just providing an inbound call service, to ensure everyone knew the different options available. Where additional capacity had been laid on, such as extra flights or larger aircraft, call centres advised customers who may have booked three or four days out, telling them if a flight with additional seats was departing sooner (Transport Select Committee, 2023j, Q41).

While the airlines reported that they had used their communication channels and apps to inform customers of the delays and cancellations and consumer rights, there were complaints that these rights were not always fully explained or provided in full. The consumer group Which? stated that passengers' testimonies posted on social media reported that in a number of instances, airlines were not fully and completely explaining passengers' rights following the NATS system failure (Transport Select Committee, 2023j). Specifically, they highlighted Ryanair (see Figure 6.1) and Jet2 (see Figure 6.2), who both posted graphics on social media containing inaccurate information about the right to rerouting.

Neither Jet2 nor Ryanair made it clear on their X posts that passengers had the option of being rerouted at the earliest opportunity or at an agreed date via any carrier or alternative transport mode.

With such significant disruption, assistance for delayed passengers was problematic. Airports' capacity to accommodate the delayed passengers was limited. Airport restaurants, cafes and retailers did not have the resources or supplies to cater for the additional demand and remain open for the extended period required. While delayed passengers might have been given food and drink vouchers as part of rights-to-care by the airline, there were significant queues of people trying to get basic drinks and food. In a number of incidences, refreshment vouchers were not accepted in many parts of the airport or were insufficient to cover the costs of a full meal (Transport Select Committee, 2023f, Q44).

The system failure did not only cause significant disruption for passengers, airlines and passengers, but it also created substantial costs. UK and EU legislation states passengers who experience cancelled flights are eligible for rights-to-care expenses. Ryanair said that they spent over £15 million to cover

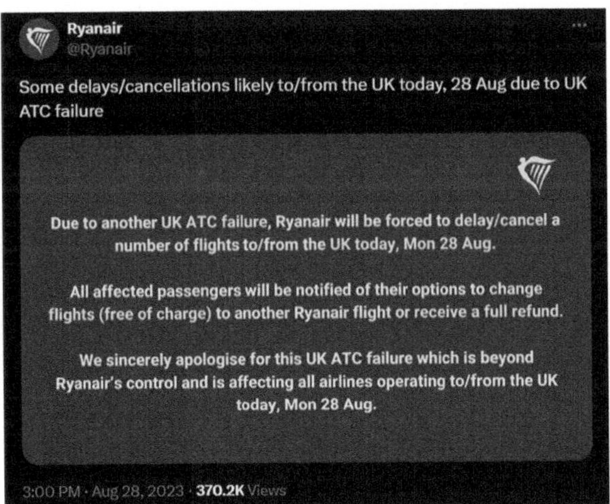

Figure 6.1 Ryanair's notice on X.

Source: X: https://x.com/Ryanair/status/1696160754787819980.

Figure 6.2 Jet 2's notice on X.

Source: X: https://x.com/jet2tweets/status/1696224590047932648.

expenses such as consumer hotel costs, meal costs and repatriation on Ryanair flights or other flights. Jonathan Hinkles, Chief Executive of Loganair, a smaller airline than Ryanair, estimated that his airline incurred right-to-care costs of around £300,000 in the space of five hours (Transport Select Committee, 2023d, Q21). Such costs do not include additional expenditures as a result of delays, such as pilots and cabin crew going out of hours, rerouting aircraft or laying on additional ones to repatriate customers.

6.5 Strategic Crisis Management and Communication Theories in Practice

Exploration of aviation disruption's crisis management and communication must be considered in light of its dual characteristics of complexity and perceived risk. The complexity arises from the intricate socio-technical interconnected subsystems and the involvement of multiple stakeholders, which require the effective interaction of human and technical components to ensure safe operation (Foster et al., 2021; Muecklich et al., 2023). Air travel is associated with risk, partly due to the complexity of its systems and the potential for human error, but also because of its susceptibility to terrorism, vulnerability to natural disasters and the challenges posed by flying across large time-space horizons (Janic, 2000; Muecklich et al., 2023).

Multiple phases are involved in any crisis or disruption management procedure, including pre-crisis, during-crisis and post-crisis, aligned with a set of functions such as crisis prevention, response and recovery (Coombs, 2021). A pillar of crisis management is the communication of crisis-related information to stakeholders, which influences their response behaviour and shapes public opinion (Coombs, 2021; Ulmer et al., 2022).

Restoration Theory is an early and influential crisis communication theory (Benoit, 1997) that focuses on the strategies used by organisations or individuals once a crisis has taken place to maintain their reputation and mitigate damage. This theory is based on the core principles that the image must be maintained, and any threat to the image should be responded to accordingly in order to mitigate any negative impact. Five broad strategies support Restoration Theory crisis communication response: Denial, Evasion of Responsibility, Reducing Offensiveness, Corrective Action and Mortification. Denial involves an organisation in outright denial that the crisis ever took place or the shifting of blame to another party to forgo any wrongdoing. This is based on the idea that if an organisation is not associated with the crisis, then it will not suffer any damage to its reputation (Coombs, 2007).

Evasion of Responsibility occurs when organisations seek, through provocation, defeasibility, lack of intention or good intent, to remove any association with causing the crisis. Provocation works on the basis that the crisis or event is a reasonable response to the provocation, because if it is not, then liability still remains with the crisis organisation. Defeasibility means that the business did not have the truth or enough information to know otherwise. Good intentions can help to rebuild the consumer's trust and mend the reputation of the organisation. Reducing Offensiveness involves removing the impact of a crisis by lessening its severity. A very similar technique is minimisation, whereby the organisation attempts to downplay the significance of the crisis to maintain confidence in the organisation (Coombs & Holladay, 2002). Benoit's Image Restoration Theory (1997) also includes differentiation, which involves placing the crisis in the context of much larger, more severe crises to distinguish the current crisis as less offensive than previous ones.

Corrective action involves taking responsibility for the issue, correcting and rectifying it, and implementing strategies to ensure the crisis never happens again. Organisations have been known to benefit from using corrective actions, as those who use them have maintained better reputations than those who have not (Claeys et al., 2010). Mortification involves admitting responsibility and expressing regret, an important strategy in maintaining the organisation's reputation. If the apology is sincere, the organisation can build trust with customers and relevant stakeholders; however, if the organisation seems insincere, it may suffer further damage to its reputation (Coombs, 2007).

To understand the response to travel disruptions, Communications and Restoration Theory can be considered in conjunction with Situational Crisis Communication Theory (Coombs, 2007). Here, the response to the crisis depends on three types of crises: victim, accidental and preventable, created by or impacting the organisation. When a crisis occurs and the organisation is not responsible but is negatively impacted, the communication response can align with the general public and stakeholders who are also victims. Denial, in this instance, is a useful strategy. Where the crisis is accidental, diminishing strategies are best used to downplay the crisis (Coombs, 2007). However, when the crisis is preventable or could have been lessened through better management, corrective action and mortification should be used to sufficiently maintain the business's reputation.

Analysing the response using a combination of Restoration Theory (Benoit, 1997) and Situational Crisis Communication Theory (Coombs, 2007) identifies the different approaches depending upon the organisation. For NATS there was a mixture of denial and evasion of responsibility (Coombs, 2007). Denial was the early response, with nearly three hours passing after the manual input of flight data began before stakeholders, such as airlines and airports, were officially informed with an email at 11:30 am (CAA, 2024).

The claim by NATS of this being an exceptional occurrence (Austin, 2023) aligns with Restoration Theory's evasion of responsibility (Benoit, 1997). According to NATS, this was a curious combination of the airline's flight plan, how Eurocontrol translated this and how the NATS systems then processed it (CAA, 2024). When interviewed by the BBC, NATS CEO Martin Rolfe stated, 'This was a one in 15 million chance. We've processed 15 million flight plans with this system up until this point and never seen this before' (Austin, 2023). Such a position could be supported because of a unique set of circumstances regarding the specific routing of the aircraft. However, Michael O'Leary, Chief Executive Officer of Ryanair, disputed such exceptional claims. He stated that such an event had happened before and that other air traffic systems 'are designed so that when they receive a duplicate flight plan like that, they reject it and deal with it manually. That is routine. It happens on a daily basis' (Transport Select Committee, 2023i, Q45).

With the possibility of a 1 in 15 million chance, NATS appears to have positioned the incident as accidental within Situational Crisis Communication Theory (Coombs, 2007). This accidental position was questioned with Michael O'Leary, Chief Executive Officer of Ryanair, stating such a situation was preventable (Transport Select Committee, 2023c, Q19). Not only do other ATC systems appear to handle flight plans with duplicate waypoints, but there is also an issue with the backup system and crisis protocols and systems used by NATS (CAA, 2024). Sean L Doyle, CEO and Chairman of British Airways, also questions this accidental position, stating, 'there are some fundamental aspects of the failure that resulted in a worse impact than otherwise might have been the case. For example, the fact that only four hours of stored flight data is available in the event of such an outage is disappointing' (Transport Select Committee, 2024).

There were also delays in identifying and rectifying the problem that undermined this accidental position. The first engineer called was working off-site and, only after exhausting remote intervention options, arrived on-site one and a half hour later to restart the system (CAA, 2024). When this did not work, following escalation protocols, the assistance of a more senior engineer was not sought for more than three hours after the initial failure (CAA, 2024). They were unfamiliar with the fault message, and there was a further delay, again due to protocols, until four hours after the initial failure before a third-party engineer from the platform provider was called. This highlights several crisis management issues concerning the availability of engineers, their knowledge of the system and the protocols applied for escalation. As Jonathan Hinkles, Chief Executive of Loganair, explained, the 'length of time directly magnified the impact of the disruption that all airlines and logically, therefore, sadly, their customers incurred on that day' (Transport Select Committee, 2023f, Q44). So while NATS may not have been responsible for the flight plan data, with different systems and protocols the level and duration of the impact could have been preventable.

The scale of the impact could also have been preventable had a contingency plan been in place for a greater number of flight plans to be processed manually. The contingency for inputting data was to revert to a manual process, referred to by Jonathan Hinkles as 'resorting to the air traffic control of the 1950s – pen, paper, and telephone' (Transport Select Committee, 2023d, Q21). With just seven manually operated terminals and a limited number of staff qualified to enter full flight plans, corrective action was restricted in what it could achieve.

Rather than evasion of responsibility, airlines appear to have engaged in Benoit's (1997) reducing offensiveness component of Restoration Theory through efforts to remove the impact of the crisis, lessen its severity and engage with rectifying the issue. They could take this stance as they were, in Situational Crisis Communication Theory (Coombs, 2007) terms, victims of the shutdown of the NATS system. Indeed, airlines have been praised for their effective responses to the travel disruption caused by the failure of the flight planning system. Rob Bishton, Chief Executive Designate, CAA commented, 'It certainly was swift. By the Wednesday, largely speaking, most of the disruption had been contained...the airlines' care, assistance and rerouting response, albeit there were exceptional cases, was good' (Transport Select Committee, 2023g, Q92).

Further support for aligning with denial and reducing offensiveness of Benoit's (1997) Restoration Theory and being a victim in Situational Crisis Communication Theory (Coombs, 2007) can be seen from some airlines' posts on social media. Ryanair's X post (Figure 6.1) included phrases such as 'another UK ATC failure,' and 'forced to delay/cancel a number of flights' along with 'we sincerely apologise for this UK ATC failure which is beyond Ryanair's control and is affecting all airlines operating to/from the UK'. Ryanair places the responsibility for the crisis with others by putting it in the context of much larger, more severe crises, i.e., affecting 'all' airlines.

A similar denial and reducing offensiveness (Coombs, 2007) approach to Ryanair was taken by Jet2, with their X post (see Figure 6.2) using terms such as 'technical fault affecting the UK's National Air Traffic Services, which has impacted all airlines' (Jet2, 2023). Jet2 has placed the disruption within the broader context, also claiming no responsibility by stating in their X post that this was 'completely outside of our control' (Jet2, 2023).

6.6 Building Future Resilience: Lessons Learnt and Strategic Recommendations

With comments from Rob Bishton, Chief Executive Designate, CAA, stating that 'performance from the airlines beyond their obligations' (Transport Select Committee, 2023h, Q98), there was a consensus that airlines responded well to the impact of the system's failure. Airlines had learnt from previous incidents such as the Icelandic Ash Cloud 2010 how to manage and respond to such issues. Despite this, there are a number of recommendations to help build future resilience in terms of legislation, communication and contingency planning.

The first recommendation is for regulations to support the enforcement of airlines' and airports' consumer duties. This could include specific provisions to apply fines when airlines fail to fulfil their legal obligations and regulatory enforcement powers similar to those for other markets such as mobile phones and energy. Such enforcement would help prevent inaccurate information (see Figures 6.1 and 6.2), including passenger rights to rerouting with other airlines (Transport Select Committee, 2023j).

The second recommendation is to enhance the mode, format and channel of communication between aviation support systems such as air traffic control with airports and airlines. Airlines believed that the delay in notification of a possible problem resulted in their losing time to help plan and cater for the disruption. While NATS stated they used normal channels through Eurocontrol, this does not appear to have been satisfactory from a crisis management and communication perspective.

The third recommendation to support future resilience and improve the response to such disruptions is to enhance and improve the methods airlines use to communicate with passengers. Specific recommendations are to ensure the information is consistent, accurate, up to date and provided in an inclusive manner to ensure that it is accessible for all passengers. While the airlines used push notifications on their apps and other communication methods, this was not always accurate, sufficient, timely or appropriate (Transport Select Committee, 2023j).

The fourth recommendation to support more impactful and timely decision-making is to appoint a single post-holder with overall management and full accountability for such incidents and their resolution (CAA, 2024). The CAA identified the NATS joint decision-making model used for incident management as a process for specific examination (CAA, 2024).

The final recommendation is to review the systemic approach adopted by NATS to ensure more robust and timely procedures for problem identification and resolution. NATS level of documentation, error identification, incorrect data isolation and correction protocols contributed to the system failure and the time taken to identify and rectify the problem (CAA, 2024). Air traffic control is an intricate socio-technical interconnected set of subsystems with multiple stakeholders. The recommendations in this chapter are to support effective response strategies that should help ensure one flight plan does not jeopardise the whole system.

References

Austin, K. (2023). Ryanair boss calls air traffic chaos report rubbish. *BBC News*. https://www.bbc.co.uk/news/business-66723586#:~:text=The%20boss%20of%20Ryanair%20has,was%20%22full%20of%20excuses%22. Accessed 8th August 2024.

BBC. (2023). Manchester couple fear they could miss dream Italian wedding. *BBC News*. https://www.bbc.co.uk/news/uk-england-manchester-66645674. Accessed 4th August 2024.

Benoit, W. L. (1997). *Accounts, excuses, and apologies: A Theory of image restoration strategies*. Albany, NY: State University of New York Press. https://doi.org/10.2307/2393739

CAA (2024). *The Civil Aviation's Independent Review of NATS (En Route) Plc's Flight Planning System Failure 2023 Interim Report (March 2024)*. https://www.caa.co.uk/our-work/publications/documents/content/cap2981/

Calder, S. (2024). The collapse of air-traffic control that caused major flight chaos and how it unfolded. *Independent*. https://www.independent.co.uk/travel/news-and-advice/flight-air-traffic-control-failure-nats-b2513127.html. Accessed 9th September 2024.

Claeys, A. Cauberghe, V., & Vyncke, P. (2010). Restoring reputations in times of crisis: An experimental study of the Situational Crisis Communication Theory and the moderating effects of locus of control. *Public Relations Review*, 36(3), 256–262. https://doi.org/10.1016/j.pubrev.2010.05.004

Coombs, W. (2007). Protecting organization reputations during a crisis: The development and application of situational crisis communication theory. *Corporate Reputation Review*, 10, 163–176. https://doi.org/10.1057/palgrave.crr.1550049

Coombs, W. (2021). *Ongoing crisis communication: Planning, managing, and responding*. New York, NY: Sage Publications.

Coombs, W., & Holladay, S.J. (2002). Helping crisis managers protect reputational assets: Initial tests of the situational crisis communication theory. *Management Communication Quarterly*, 16(2), 165–186. https://doi.org/10.1177/0893318802237233

Foster, C. J., Plant, K. L., & Stanton, N. A. (2021). A very temporary operating instruction: Uncovering emergence and adaptation in air traffic control. *Reliability Engineering & System Safety*, 208, 107386. https://doi.org/10.1016/j.ress.2020.107386

IATA (2024). Global outlook for air transport deep change. International Air Transport Association Montreal. https://www.iata.org/en/iata-repository/publications/economic-reports/global-outlook-for-air-transport-june-2024-report/

Janic, M. (2000). An assessment of risk and safety in civil aviation. *Journal of Air Transport Management*, 6(1), 43–50. https://doi.org/10.1016/s0969-6997(99)00021-6

Jet2 [@Jet2] (2023). Important notice 28th August 2023. X post. https://x.com/jet2tweets/status/1696224590047932648

Muecklich, N., Sikora, I., Paraskevas, A., & Padhra, A. (2023). Safety and reliability in aviation–A systematic scoping review of normal accident theory, high-reliability theory, and resilience engineering in aviation. *Safety science*, 162, 106097. https://doi.org/10.1016/j.ssci.2023.106097

Murphy, G., & Efthymiou, M. (2017). Aviation safety regulation in the multi-stakeholder environment of an airport. *Journal of Air Transport Studies*, 8(2), 1–26. https://doi.org/10.38008/jats.v8i2.30

Race, M. (2023). Families worried about expenses after flights axed. *BBC News*. https://www.bbc.co.uk/news/business-66657176. Accessed 8th September 2024.

Transport Select Committee (18th October 2023a). Air traffic control disruption, Oral Evidence, HC1849. Q17 https://committees.parliament.uk/event/18993/formal-meeting-oral-evidence-session/

Transport Select Committee (18th October 2023b). Air traffic control disruption, Oral Evidence, HC1849. Q18 https://committees.parliament.uk/event/18993/formal-meeting-oral-evidence-session/

Transport Select Committee (18th October 2023c). Air traffic control disruption, Oral Evidence, HC1849. Q19 https://committees.parliament.uk/event/18993/formal-meeting-oral-evidence-session/

Transport Select Committee (18th October 2023d). Air traffic control disruption, Oral Evidence, HC1849. Q21 https://committees.parliament.uk/event/18993/formal-meeting-oral-evidence-session/

Transport Select Committee (18th October 2023e). Air traffic control disruption, Oral Evidence, HC1849. Q22 https://committees.parliament.uk/event/18993/formal-meeting-oral-evidence-session/

Transport Select Committee (18th October 2023f). Air traffic control disruption, Oral Evidence, HC1849. Q44 https://committees.parliament.uk/event/18993/formal-meeting-oral-evidence-session/

Transport Select Committee (18th October 2023g). Air traffic control disruption, Oral Evidence, HC1849. Q92 https://committees.parliament.uk/event/18993/formal-meeting-oral-evidence-session/

Transport Select Committee (18th October 2023h). Air traffic control disruption, Oral Evidence, HC1849. Q98 https://committees.parliament.uk/event/18993/formal-meeting-oral-evidence-session/

Transport Select Committee (2023i). Air traffic control disruption, Oral Evidence, 18th October 2023, HC1849. Q45 https://committees.parliament.uk/event/18993/formal-meeting-oral-evidence-session/

Transport Select Committee (2023j). Session on NATS system failure, written evidence, submitted by which? https://committees.parliament.uk/publications/41703/documents/206556/default/

Transport Select Committee (2024). Written evidence: air traffic control disruption hearing. Submitted by British Airways https://committees.parliament.uk/publications/42114/documents/209387/default/

Ulmer, R. R., Sellnow, T. L., & Seeger, M. W. (2022). *Effective crisis communication: Moving from crisis to opportunity*. New York, NY: Sage Publications.

Wilson, R. (2023). Air traffic control: Woman misses heart transplant check-up over delay. *BBC News*. https://www.bbc.co.uk/news/uk-northern-ireland-66640003. Accessed 8th September 2024.

7 Navigating Post-Pandemic Destination Marketing Communication

A Synthesis of Crisis Communication and Tourism Disaster Management Theories

Shweta Singh, Annmarie Nicely, Jonathon Day, Liping A. Cai and Xinran Lehto

7.1 Introduction

The disruption caused by the pandemic has underscored the need for robust and strategic communication to restore traveller confidence and rebuild the tourism sector. Despite this urgency, a dearth of empirical research exists to guide tourism marketers in crafting post-disaster communication, necessitating an often ad-hoc approach that may lack efficacy. Addressing this gap is the central objective of this chapter.

This chapter begins with an exploration of crisis communication and resilience theories drawn from the organizational crisis communication and behavioural science literature to understand their applicability to the unique challenges destination communities face during a pandemic. Finally, the chapter presents a set of guidelines strategically assembled to assist destinations in conceptualizing post-pandemic destination marketing communication. These guidelines are designed to help tourism marketers develop effective communication strategies that can enhance visitor confidence, promote safe travel practices, and ultimately support the recovery and resilience of the tourism sector. This chapter aims to provide valuable insights and tools for tourism professionals navigating the complexities of post-pandemic recovery by bridging the gap between theoretical knowledge and practical application.

7.2 Disaster and Crisis Defined

The terms crisis and disaster are used interchangeably in tourism literature. Faulkner (2001) offered a distinction between the two. He defined crisis as "a situation where the root cause of an event is, to some extent, self-inflicted through such problems as inept management structures and practices or a failure to adapt to change" and disaster as "a situation where an enterprise or a destination is confronted with sudden, unpredictable catastrophic changes over which it has little control" (p. 136). Coles (2004, p. 175) added that "[disasters] take a variety of forms from natural landscape disasters to episodes of

DOI: 10.4324/9781032720555-8

scarcity, disease, wars, terrorist atrocities, and political instability." Some researchers prefer to use the term disaster in the context of destinations, as it captures an element of unpredictability and lack of control (e.g., Walters & Mair, 2012). Thus, this chapter also employed the term disaster for pandemics, as destination communities, tourism organizations, and hospitality businesses have limited control over their unpredictable effects.

7.3 Disaster and Crisis Communication

The tourism industry is particularly vulnerable to disasters, as it relies heavily on visitor perceptions (Malhotra & Venkatesh, 2009). In addition, Coombs and Holladay (2011) argued that crisis communication is critical in "protecting the organizational reputation, … and increasing potential supportive behavior" (p. 223). Coombs and Holladay (2011, p. 20) defined crisis communication as "the collection, processing, and dissemination of information required to address a crisis."

Moreover, crisis communication is multifaceted in the context of the tourism industry. "The characteristics of tourism create an integrated and open system which is impacted by external factors, making it vulnerable to crises and disasters" (Ritchie, 2009, p. 13). The tourism industry, encompassing many public and private sector organizations, ranging from airlines, hotels, and cruises to transport services, restaurants, entertainment venues, and attractions, forms a complex network. In contrast to organizational crises, disasters impacting tourism extend beyond isolated entities, jeopardizing entire destinations and compromising the safety of residents and visitors. This complexity is heightened during pandemics, where repercussions manifest at local, national, and global levels of tourism activity. Thus, destination communities, tourism organizations, and hospitality businesses must take appropriate disaster response measures to ensure visitor safety and maintain positive perceptions of destinations (Barbe & Pennington-Gray, 2020).

7.4 Theoretical Foundation for Disaster Communication

Researchers have ascribed great importance to effective crisis communication, which has led to the growth of theories and models for organizational crisis response (Coombs & Holladay, 2012). As some of the most noteworthy fundamental crisis communication theories originated from the organizational crisis communication literature, this section outlines the applicability of these theories for destination communities, tourism organizations, and hospitality businesses in the context of a global pandemic such as COVID-19.

7.4.1 Image Restoration or Repair Theory

Image restoration theory states that organizations have a reputation or an image that must be protected from attack. For Benoit (1997), an attack on one's image has two components: (1) an offensive act and (2) an accusation of responsibility for the act. There is no reputational threat if either of these elements is absent (Coombs & Holladay, 2012).

Benoit (1995) outlined five image restoration strategies, emphasizing that communication shapes stakeholder perception during crises. These strategies are denial, evading responsibility, reducing offensiveness, corrective action, and mortification. Denial involves refusing responsibility, with strategies like simple denial (disavowing involvement) and blame-shifting. These strategies may not suit prolonged disasters like pandemics, which cannot be denied. Burke (1970) termed blame-shifting as victimage, where the accused claims to be a victim. Destinations are also the victims since the industry has endured substantial losses from COVID-19 due to global bans on international travel.

The strategies of evading responsibility may not suit the prescribed context since there is no accusation on destination communities, tourism organizations, and hospitality businesses. However, by communicating that the destination is also a victim (*victimage*) and has no direct control (*defeasibility*) over the negative effects of the COVID-19 pandemic, destinations may elicit sympathy and foster solidarity with visitors (Walters & Mair, 2012).

The third category of response strategies is reducing offensiveness, with six variants: bolstering, minimization, differentiation, transcendence, attacking the accuser, and compensation (Benoit, 1995). Bolstering involves highlighting positive attributes or past positive acts to offset negative perceptions. Minimization downplays the crisis, which may be ineffective and risky during a pandemic, potentially leading to a loss of trust. Differentiation separates the wrongful act from more offensive acts. Transcendence places the negative act in a different, more favourable context. Attacking the accuser diverts attention from the offence. Compensation communicates a desire to make amends. In the context of a pandemic, the strategies of bolstering (highlighting good deeds), transcendence (framing the negative as a positive), and compensation may benefit destinations.

The fourth strategy, corrective action, involves the accused vowing to fix the problem. This can mean restoring the situation to its previous state. The fifth strategy, mortification, means confessing and seeking forgiveness. Both strategies are used when the accused cannot avoid or downplay their responsibility.

7.4.2 Attribution Theory

Heider and Simmel (1944) pioneered the attribution theory, which seeks to help people make sense of the world by identifying causes for events they experience. Even though destination communities, tourism organizations, and hospitality businesses hold no direct responsibility for the occurrence of pandemics, they are still judged by their responses and reactions to these disasters (Faulkner, 2001). Visitors expect them to ensure their safety during and after a pandemic. Situation crisis communication theory (SCCT) translated and extended ideas from attribution theory to a crisis context.

7.4.3 *Situational Crisis Communication Theory*

The next significant advancement in crisis communication literature was W. Timothy Coombs' (1995) SCCT, which involves a two-step process for assessing potential judgements of crisis responsibility. First, the basic crisis type is evaluated, categorizing crises as (1) victim, where external forces like terrorism, natural disasters, or pandemics attack the organization; (2) accidental, resulting from errors; and (3) preventable, where management knowingly places stakeholders at risk. Victim crises generally attract minimal responsibility attribution. The second step is identifying intensifying factors, which can make organizations perceived poorly for even victim and accidental crises. Such factors include a bad reputation and a history of crises (Coombs & Holladay, 1996). A global pandemic typically falls into the victim category, with destination communities, tourism organizations, and hospitality businesses also seen as victims. However, factors like a negative image, perceived safety inadequacies, or a poor crisis management history can make visitors blame destination communities for the disaster.

Coombs and Holladay (2011) identified two primary roles of crisis communication: (1) managing information and (2) managing meaning. Managing information involves gathering and sharing crisis-related details while managing meaning focuses on shaping perceptions of the crisis and the organization. According to SCCT, organizations should first protect the public from harm by instructing and adjusting information before managing perceptions of their responsibility in a crisis (Coombs, 2007). During the H1N1 crisis, Kim and Liu (2012) observed that government organizations prioritized instructing information (e.g., how to stay safe), whereas corporate organizations emphasized reputation management. They suggested that destination communities, tourism organizations, and hospitality businesses risk harming their long-term reputations if they focus solely on reputation repair during health disasters.

Coombs (2008) further proposed ten general communication strategies within three responsibility postures: (1) deny, which includes attack, denial, and scapegoat; (2) diminish, includes the excuse and justification approaches; and (3) deal, which includes the concern, compensation, regret, and apology strategies. In the deny posture, organizations either refute crisis allegations or refuse to acknowledge them. Scapegoating shifts the blame on others.

When denial is not credible, organizations can manage crises by compensating victims financially or showing concern. An apology is used when organizations express regret for a crisis. SCCT further recommends reinforcement strategies that include bolstering (i.e., highlighting past good deeds), ingratiation (i.e., praising stakeholders), and victimage (i.e., declaring themselves as victims) (Coombs, 2007). In the context of health disasters, Liu (2010) identified endorsement as an additional reinforcing strategy in which organizations employ third-party support in their crisis response.

Kim and Liu (2012) tested SCCT during the 2009 H1N1 influenza disaster, finding that reputation-focused responses were ineffective, while reinforcement strategies were more suitable for destinations. They recommended bolstering, ingratiation, and victimage to build positive stakeholder connections. Barbe and Pennington-Gray (2020) observed similar results in Orlando hotels' post-disaster communications after the 2016 Pulse Nightclub shooting. Kim and Liu (2012) further discovered enhancing (focusing on current good deeds) and transferring (supporting a credible third party's response) as fitting strategies.

7.5 The Multi-step Model of Altering Place Image

The multi-step model of altering place image by Avraham and Ketter (2008a) is a pioneering framework for destination image restoration after a disaster. Avraham and Ketter (2008a) argue that existing image repair models, which focus solely on organizational contexts, are not applicable to destinations. Unlike organizational models, this model views destination image repair as a long-term process where destinations cannot shift blame or accept responsibility.

The model's first step involves analysing crisis characteristics, target audience, and place (CAP). Crisis characteristics include scale, duration, stage, threat level, damage, and casualties. Audience characteristics include proximity to disaster location, knowledge about the destination, type and size of the target group, sources of information, and shared values. Place characteristics encompass power and status, resource availability, location, lifestyle, and leadership at the destination. Based on the outcomes of a CAP analysis, marketers can define their campaign objectives and timing. Thus, the model recommends that post-pandemic destination marketing messages be finalized according to the characteristics of the pandemic, the target audience, and the location.

7.6 Communication Theory of Resilience

The communication theory of resilience (CTR) by Patrice Buzzanell (2010) focuses on communication processes that may result in resilience. CTR uses Richardson's (2002, p. 309) definition of resilience as "the process of reintegrating from disruptions in life" and presents it as a desired outcome of disaster communication.

CTR includes five communication processes that facilitate resilience: crafting normalcy, affirming identity anchors, using and maintaining communication networks, applying alternative logic, and foregrounding productive action. Crafting normalcy concerns the notion of reverting "back to normal" and discusses using communication to recapture a pre-disaster sense of life (Buzzanell, 2010). It is believed that things might not return to how they were before the COVID-19 pandemic (Sigala, 2020), and post-disaster destination communication for long-term recovery must complement the new reality (Scott et al., 2008). Avraham and Ketter (2008b) recommended using messages that focus on the recovering landscape to enhance visitors' interest.

Maintaining and using communication networks involves leveraging existing connections and networks to recover from a disaster. Affirming identity anchors describe how individuals or organizations may use communication to regain and maintain a pre-disaster sense of self. Identity anchors are an enduring set of identity narratives that individuals use to explain who they are and how they relate to those around them (Buzzanell, 2010). The COVID-19 "we are in this together" message exemplifies this strategy.

Putting alternative logic to work requires reframing the perception of adversity (Buzzanell, 2010). Strategies of transcendence (Benoit, 1995) and reframing a pandemic as an opportunity to reimagine tourism at a destination are examples of applying alternative logic in disaster communication. Finally, foregrounding productive action involves focusing on the positive aspects of a given situation while managing the negatives, for example, offering evidence of successful pandemic management at a destination in marketing communication.

7.7 Crafting Evidence-based Post-pandemic Destination Marketing Communication

Crisis communication in tourism is multifaceted. While crisis communication literature focuses on one organization, tourism disasters affect multiple organizations in the tourism system (Barbe & Pennington-Gray, 2020). Thus, post-pandemic communication goes beyond stakeholders' perceptions or destination image repair. As Oliveira and Huertas (2019) explained, the role of marketing communication for destination recovery after a disaster expands to conveying the message of visitor safety in the future and promoting tourism renewal.

7.8 Key Considerations for Designing Post-pandemic Destination Marketing Communication

This chapter reviewed various crisis communication theories that offered the following key considerations for designing destination marketing communication for post-pandemic recovery.

7.8.1 Attribution of Responsibility

The crisis communication theories discussed in this chapter have two key assumptions; there must be an offensive act, and destination communities, tourism organizations, and hospitality businesses are held accountable for the offence. IRT (Benoit, 1995) posits no reputational threat without blame. SCCT (Coombs, 1995) proposed communication strategies based on attributions of responsibility for crises and framed a pandemic as a victim crisis. Destination communities, tourism organizations, and hospitality businesses are framed as victims of the pandemic, producing minimal attribution. However, situational factors like a prior bad reputation and a history of poorly handling similar crises may intensify the attribution. Thus, responsibility in a pandemic context includes management of the pandemic at the destination.

Further, researchers have discussed cognitive bias associated with attribution (Jackson, 2019). The attribution theory (Heider, 1958) conceptualized this attribution bias as a self-protective strategy where visitors fault destination communities, tourist organizations, and hospitality businesses for adversity and inadequate pandemic response. Thereby, destinations are not free from responsibility and would be held accountable by visitors for ensuring their safety while offering a full range of travel experiences (Oliveira & Huertas, 2019). Destinations must establish and communicate corrective measures to avoid risking their reputation (Zizka et al., 2021).

7.8.2 Salient Audience

In crisis communication, the perceptions of the salient audience on destinations' obligations are more important than their actual role (Benoit, 1997). The key question is not if the destination is safe but whether visitors believe it to be safe. As Avraham and Ketter (2008a) suggested, the target audience is central to crisis communication strategies. A destination has numerous stakeholders, including governments, business partners, competitors, local communities, and visitors, each with distinct interests. Visitors can be further segregated into several target markets, and a destination may have more than one target market of interest. Theories suggest destinations identify the most salient audience and design communication that fits their characteristics.

7.8.3 Nature of the Crises

The nature of the crisis is another key consideration in designing post-disaster communication. SCCT suggests long-lasting crises require greater attention and more visitor-focused responses (Coombs, 2015). The multi-step model of altering place image (Avraham & Ketter, 2008a) advises analysing crisis characteristics that may include scale, duration, stage of the crisis, threat level, damage, and casualties. In the context of a pandemic, the number of confirmed cases, casualties, state of vaccination, recovery, and travel restrictions at the destination country may also be considered. Zhang et al. (2022) agreed that the type of crisis must align with the source of communication to positively influence visitors' information processing, perception of safety, and travel intentions after COVID-19.

7.8.4 Instructing and Adjusting Information

The Situational Crisis Communication Theory suggests instructing and adjusting information as the first step in crisis response. Destinations should instruct visitors on how to protect themselves physically. It may include symptoms and precautionary measures of COVID-19, safety protocols, and availability of medical resources at the destination (Coombs, 2015; Mele et al., 2023). On the other hand, adjust information to help visitors handle the psychological consequences of the pandemic, including expressing sympathy, information about the pandemic, counselling, and corrective actions. Corrective actions explain how the destination managed the pandemic and what is being done to prepare for future disasters (Coombs, 2015). Expression of sympathy helps subdue anger, while counselling can manage anxiety caused by the pandemic. Table 7.1

Table 7.1 Key considerations for designing post-pandemic destination marketing communication

Key Considerations	Details
Attribution of responsibility	Destinations may be blamed for • Inadequate response to the pandemic • Failing to provide necessary information • Failing to protect visitors or ensure visitor safety • Failing to offer a full range of travel experiences
Characteristics of the pandemic	Analyse the characteristics of the pandemic before finalizing post-disaster communication. Pandemic characteristics may include • Scale of pandemic • Duration of pandemic • Stage of the pandemic • Level of threat • Level of damage • Casualties caused • State of recovery at the destination • State of travel bans, vaccination, or other regulatory mandates
Characteristics of the salient audience	Analyse the characteristics of the target audience before finalizing post-disaster communication. Audience characteristics may include • Proximity to the crisis location • Type and size of the target group • Knowledge about the destination • Source of information • Shared values • Demographic characteristics • Political views • Pandemic knowledge • Perception of pandemic risk • Exposure to infectious diseases • Travel history
Intensifying factors	• Prior bad reputation • History of poorly handling similar crises • Poor management of the pandemic at the destination
Instructing information	Offer information for all visitors to protect them from physical harm. It may include • Symptoms and precautionary measures • Safety protocols • Availability of medical resources at the destination
Adjusting information	Offer information for all visitors to protect them from physiological harm. It may include • Expressing sympathy • Information about the pandemic • Counselling • How the destination managed the pandemic • What is being done to prepare for future disasters

summarizes these key considerations for designing post-pandemic crisis communication for destination recovery.

7.8.5 Strategies of Post-pandemic Destination Marketing Communication

Based on the review of prevailing theories of crisis communication in this chapter, a set of fitting post-pandemic marketing communication strategies are identified as follows.

7.8.5.1 Safety

Before COVID-19, safety was not emphasized in crisis communication. The pandemic highlighted visitors' safety concerns, prompting destinations to incorporate safety assurances in their marketing communications (Ketter & Avraham, 2021). Singh et al. (2022) examined academic and industry experts using a three-round Delphi technique (Singh et al., 2021) right after the COVID-19 pandemic. They recommended emphasizing safety in marketing communication to boost visitors' confidence in international travel. Mele et al. (2023) also found that Destination Management Organizations (DMOs) promoted safety information in their marketing communications during the COVID-19 pandemic.

7.8.5.2 Pandemic Information

Destinations should provide comprehensive health and safety information, managing visitors' perceptions and information needs (Coombs, 2008, 2015). Clear communication about COVID-19 safety protocols and status can reduce visitors' anxiety and anger, influencing travel intentions positively (Singh et al., 2022; Zhu & Deng, 2020).

7.8.5.3 Praise Stakeholder

Praising stakeholders (ingratiation) can enhance positive perceptions and build goodwill (Coombs, 2007). Kwok et al. (2021) used SCCT to review social media messages from eight large hotel chains between January and mid-June 2020. They found that ingratiation was the most frequently used communication strategy during COVID-19.

7.8.5.4 Victimage

Positioning destinations as victims of the pandemic can garner sympathy and reduce responsibility attribution (Benoit, 2015; Kim & Liu, 2012). Additionally, Kim and Liu (2012) found victimage effective in building positive connections with stakeholders after a health disaster. Ketter and Avraham (2021) also noted the widespread use of the words "we," "us," and "together" in destination marketing campaigns after COVID-19. "We are in this together" messages indirectly communicate that the destinations are also victims of the pandemic – the same as the visitors.

7.8.5.5 Highlight Current Good Deeds

Focusing on current positive actions (enhancing) can help maintain and repair a destination's reputation (Benoit, 1997; Coombs, 2007). The CTR (Buzzanell, 2010) supports these recommendations by advocating for foregrounding productive actions to create resilience. Furthermore, highlighting positive information about the destination may develop positive feelings among visitors, reducing anxiety and promoting positive behavioural outcomes (Coombs, 2007). Academic and industry leaders also suggested showcasing "what is good at the destination" and "proof of managing the pandemic" to attract visitors after COVID-19 (Singh et al., 2022). Leung et al. (2022) confirmed that this strategy improves visitor engagement and visit intentions after COVID-19.

7.8.5.6 Support a Credible Third Party's Response

The CTR (Buzzanell, 2010) suggested relying on established networks to develop resilience. Kim and Liu (2012) also identified endorsement as an effective post-disaster communication strategy. Endorsing credible third-party responses can transfer credibility to the destination, positively influencing visitors' perceptions (Heider, 1958; Kim & Liu, 2012). Partnerships with various stakeholders were recognized as a key marketing strategy in recovering Singapore tourism after the 2003 SARS outbreak (Beirman, 2006). Recently, Huang et al. (2024) suggested that destinations ask social media influencers to visit and share their experiences to boost visitors' confidence in the destination after a disaster.

7.8.5.7 Crafting Normalcy

Restoring a sense of normalcy through marketing communication can help visitors feel more at ease and willing to travel (Buzzanell, 2010). The Delphi study by Singh et al. (2022) also recommended highlighting the destination's return to regular activities in post-COVID-19 marketing communication.

7.8.5.8 Use Identity Anchors

Reaffirming a destination's core identity and brand can aid in recovery by reminding visitors of their pre-disaster connection (Buzzanell, 2010; Ketter & Avraham, 2021). Similarly, experts proposed recapping the destination's unique offerings in post-COVID-19 marketing communication to encourage travel (Mele et al., 2023; Singh et al., 2022).

7.8.5.9 Apply Alternative Logic

Reframing the pandemic as an opportunity for positive change can attract visitors by creating a sense of optimism and control (Avraham, 2016; Buzzanell, 2010). Experts have recommended framing the pandemic in a positive light to influence travel intentions (Benoit, 1995; Ketter & Avraham, 2021).

7.8.5.10 *Appeal to Visitors' Sentiments*

Emotional appeals, including solidarity and empathy, can effectively influence travel intentions by reinforcing a sense of shared experience and support (Huang et al., 2024; Ketter & Avraham, 2021; Singh et al., 2022). In the context of the COVID-19 pandemic, Xie et al. (2021) suggested that empathy may induce travel intentions. Ketter and Avraham (2021) also identified that post-COVID-19 marketing messages included the emotions of yearning and nostalgia to remind visitors about the destination. Table 7.2 summarizes the theory-based communication strategies discussed.

Table 7.2 Overview of theory-based strategies for designing post-pandemic destination marketing communication

Type of Post-pandemic Marketing Message	Key Message Characteristics	Targeted Outcome
Select one or more of the following marketing messages per the desired outcome.		
Safety	The destination is safe and prepared	• Reduce anxiety and anger • Increase travel intentions • Rebuild visitor confidence
Accurate pandemic information	Instructing and adjusting information	• Reduce anxiety and anger • Protect reputation • Build confidence • Increase travel intentions
Highlight past good deeds	Highlight the destination's positive characteristics	• Protect reputation • Increase travel intentions
Praise stakeholders	Praise stakeholders' good deeds	• Build relationship • Increase travel intentions
Victimage	Being a victim	• Gain sympathy & support • Increase travel intentions
Highlight current good deeds	Show proof of managing the pandemic	• Protect reputation • Reduce anxiety • Increase travel intentions
Support a credible third party's response	Utilize third-party's messages in destination's marketing	• Gain credibility • Increase travel intentions
Crafting normalcy	Back to normalBusiness as usual	• Reduce anxiety • Increase travel intentions
Use identity anchors	Reminding visitors of their past connection with the destination	• Build relationship • Increase travel intentions
Apply alternative logic	Reframe the pandemic as an opportunity	• Reduce anxiety • Increase travel intentions
Foreground positive actions	Showcase what is new at the destination Show evidence of management of the pandemic	• Protect reputation • Reduce anxiety • Increase travel intentions
Appeal to visitors' sentiments	Altruistic appeal for visitors' support in recovery	• Gain sympathy & support • Increase travel intentions

7.9 Conclusion

This chapter has provided a comprehensive examination of strategic communication approaches necessary for the post-pandemic recovery of the tourism sector. The exploration of crisis communication and resilience theories, such as Image Restoration Theory, Attribution Theory, Situational Crisis Communication Theory, and the multi-step model of altering place image, demonstrated their applicability in addressing the unique challenges that tourism marketers face in the wake of global disruptions.

The chapter concluded with a set of guidelines designed to assist tourism marketers in crafting effective post-pandemic communication. Key recommendations include emphasizing visitor safety, providing transparent and comprehensive pandemic-related information, praising stakeholders, positioning destinations as victims of the pandemic, and highlighting positive actions taken during the crisis. Additionally, supporting credible third-party responses, restoring a sense of normalcy, reaffirming destination identity, applying alternative logic, and appealing to visitors' sentiments were identified as critical strategies to enhance visitor confidence and promote safe travel practices. Implementing these strategies can help destination communities, tourism organizations, and hospitality businesses build resilience, foster visitor trust, and accelerate the recovery process in the post-pandemic evolving landscape of global tourism.

References

Avraham, E. (2016). Destination marketing and image repair during tourism crises: The case of Egypt. *Journal of Hospitality and Tourism Management, 28*, 41–48. https://doi.org/10.1016/j.jhtm.2016.04.004

Avraham, E., & Ketter, E. (2008a). The multi-step model for altering place image. In E. Avraham & E. Ketter (Eds.), *Media strategies for marketing places in crisis* (1st ed., pp. 187–203). Routledge. https://doi.org/10.1016/B978-0-7506-8452-1.50016-0

Avraham, E., & Ketter, E. (2008b). Will we be safe there? Analyzing strategies for altering unsafe place images. *Place Branding and Public Diplomacy, 4*(3), 196–204. https://doi.org/10.1057/pb.2008.10

Barbe, D., & Pennington-Gray, L. (2020). Social media and crisis communication in tourism and hospitality. In Xiang Z., Fuchs M., Gretzel U., & Wolfram H. (Eds.), *Handbook of e-Tourism* (pp. 1–27). Springer. https://doi.org/10.1007/978-3-030-05324-6_130-1

Beirman, D. (2006). A comparative assessment of three Southeast Asian tourism recovery campaigns: Singapore roars: post SARS 2003, Bali post-the October 12, 2002 bombing, and WOW Philippines 2003. In Y. Mansfeld & A. Pizam (Eds.), *Tourism, security and safety* (1st ed.). Routledge.

Benoit, W. (2015). Image repair theory in the context of strategic communication. In D. Holtzhausen & A. Zerfass (Eds.), *Routledge handbook of strategic communication* (pp. 303–311). Routledge.

Benoit, W. L. (1995). *Accounts, excuses and apologies: A theory of image restoration strategies*. State University of New York Press.

Benoit, W. L. (1997). Image repair discourse and crisis communication. *Public Relations Review, 23*(2), 177–186. https://doi.org/10.1016/S0363-8111(97)90023-0

Burke, K. (1970). *The rhetoric of religion: Studies in logology* (Vol. 188). University of California Press.

Buzzanell, P. M. (2010). Resilience: Talking, resisting, and imagining new normalcies into being. *Journal of Communication, 60*(1), 1–14. https://doi.org/10.1111/j.1460-2466.2009.01469.x

Coles, T. (2004). A local reading of a global disaster. *Journal of Travel & Tourism Marketing, 15*(2–3), 173–197. https://doi.org/10.1300/J073v15n02_10

Coombs, W. T. (1995). Choosing the right words: The development of guidelines for the selection of the "appropriate" crisis-response strategies. *Management Communication Quarterly, 8*(4), 447–476. https://doi.org/10.1177/0893318995008004003

Coombs, W. T. (2007). Protecting organization reputations during a crisis: The development and application of situational crisis communication theory. *Corporate reputation review, 10*(3), 163–176.

Coombs, W. T. (2008). Conceptualizing crisis communication. In R. L. Heath & H. D. O'Hair (Eds.), *Handbook of risk and crisis communication* (1st ed., p. 696). Taylor & Francis Group. http://ebookcentral.proquest.com/lib/purdue/detail.action?docID=446682

Coombs, W. T. (2015). The value of communication during a crisis: Insights from strategic communication research. *Business Horizons, 58*(2), 141–148. https://doi.org/10.1016/j.bushor.2014.10.003

Coombs, W. T., & Holladay, J. S. (2012). The paracrisis: The challenges created by publicly managing crisis prevention. *Public Relations Review, 38*(3), 408–415. https://doi.org/10.1016/j.pubrev.2012.04.004

Coombs, W. T., & Holladay, S. J. (1996). Communication and attributions in a crisis: An experimental study in crisis communication. *Journal of Public Relations Research, 8*(4), 279–295. https://doi.org/10.1207/s1532754xjprr0804_04

Coombs, W. T., & Holladay, S. J. (2011). *The handbook of crisis communication* (Vol. 22). John Wiley & Sons. https://books.google.com/books?id=KMvtvG4zWaYC

Faulkner, B. (2001). Towards a framework for tourism disaster management. *Tourism Management, 22*(2), 135–147. https://doi.org/10.1016/S0261-5177(00)00048-0

Heider, F. (1958). *The psychology of interpersonal relations*. Lawrence Erlbaum Associates.

Heider, F., & Simmel, M. (1944). An experimental study of apparent behavior. *The American Journal of Psychology, 57*(2), 243–259. https://doi.org/10.2307/1416950

Huang, M., Wang, K., Liu, Y., & Xu, S. (2024). The impact of post-disaster communication on destination visiting intention. *Journal of Hospitality and Tourism Insights, 7*(2), 783–799. https://doi.org/10.1108/JHTI-10-2022-0519

Jackson, M. (2019). Utilizing attribution theory to develop new insights into tourism experiences. *Journal of Hospitality and Tourism Management, 38*, 176–183. https://doi.org/10.1016/j.jhtm.2018.04.007

Ketter, E., & Avraham, E. (2021). #StayHome today so we can #TravelTomorrow: Tourism destinations' digital marketing strategies during the Covid-19 pandemic. *Journal of Travel & Tourism Marketing, 38*(8), 819–832. https://doi.org/10.1080/10548408.2021.1921670

Kim, S., & Liu, B. F. (2012). Are all crises opportunities? A comparison of how corporate and government organizations responded to the 2009 flu pandemic. *Journal of Public Relations Research, 24*(1), 69–85. https://doi.org/10.1080/1062726X.2012.626136

Kwok, L., Lee, J., & Han, S. H. (2021). Crisis communication on social media: What types of COVID-19 messages get the attention? *Cornell Hospitality Quarterly, 63*(4), 528–543. https://doi.org/10.1177/19389655211028143

Leung, X. Y., Wu, L., & Sun, J. (2022). Exploring secondary crisis response strategies for airlines experiencing low-responsibility crises: An extension of the situational crisis communication theory. *Journal of Travel Research, 62*(4), 878–892. https://doi.org/10.1177/00472875221095210

Liu, B. F. (2010). Effective public relations in racially-charged crises: Not black or white. In W. T. Coombs & S. J. Holladay (Eds.), *The handbook of crisis communication* (Vol. 22, pp. 335–358). Wiley-Blackwell.

Malhotra, R., & Venkatesh, U. (2009). Pre-crisis period planning: Lessons for hospitality and tourism. *Worldwide Hospitality and Tourism Themes, 1*(1), 66–74. https://doi.org/10.1108/17554210910949896

Mele, E., Filieri, R., & De Carlo, M. (2023). Pictures of a crisis. Destination marketing organizations' Instagram communication before and during a global health crisis. *Journal of Business Research, 163*, 113931. https://doi.org/10.1016/j.jbusres.2023.113931

Oliveira, A., & Huertas, A. (2019). How do destinations use Twitter to recover their images after a terrorist attack? *Journal of Destination Marketing & Management, 12*, 46–54. https://doi.org/10.1016/j.jdmm.2019.03.002

Richardson, G. E. (2002). The metatheory of resilience and resiliency. *Journal of Clinical Psychology, 58*(3), 307–321. https://doi.org/10.1002/jclp.10020

Ritchie, B. W. (2009). *Crisis and disaster management for tourism.* Channel View Publications.

Scott, N., Laws, E., & Prideaux, B. (2008). Tourism crises and marketing recovery strategies. *Journal of Travel & Tourism Marketing, 23*(2–4), 1–13. https://doi.org/10.1300/J073v23n02_01

Sigala, M. (2020). Tourism and COVID-19: Impacts and implications for advancing and resetting industry and research. *Journal of Business Research, 117*, 312–321. https://doi.org/10.1016/j.jbusres.2020.06.015

Singh, S., Nicely, A., Cai, L., & Day, J. (2021). Visitor (tourist) harassment research: Delphi panels. *Research Methods.* https://docs.lib.purdue.edu/htmvh

Singh, S., Nicely, A., Day, J., & Cai, L. A. (2022). Marketing messages for post-pandemic destination recovery – A Delphi study. *Journal of Destination Marketing & Management, 23*, 100676. https://doi.org/10.1016/j.jdmm.2021.100676

Walters, G., & Mair, J. (2012). The effectiveness of post-disaster recovery marketing messages: The case of the 2009 Australian bushfires. *Journal of Travel & Tourism Marketing, 29*(1), 87–103. https://doi.org/10.1080/10548408.2012.638565

Xie, C., Zhang, J., Morrison, A. M., & Coca-Stefaniak, J. A. (2021). The effects of risk message frames on post-pandemic travel intentions: The moderation of empathy and perceived waiting time. *Current Issues in Tourism*, 1–20. https://doi.org/10.1080/13683500.2021.1881052

Zhang, J., Xie, C., Chen, Y., Dai, Y.-D., & Yi-Jun, W. (2022). The matching effect of destinations' crisis communication. *Journal of Travel Research, 62*, 004728752110675. https://doi.org/10.1177/00472875211067548

Zhu, H., & Deng, F. (2020). How to influence rural tourism intention by risk knowledge during covid-19 containment in China: Mediating role of risk perception and attitude. *International Journal of Environmental Research and Public Health, 17*(10), 3514. https://doi.org/10.3390/ijerph17103514

Zizka, L., Chen, M.-M., Zhang, E., & Favre, A. (2021). *Hear no virus, see no virus, speak no virus: Swiss hotels' online communication regarding Coronavirus.* Springer. https://doi.org/10.1007/978-3-030-65785-7_43

8 Travel disruptions in travel agencies

Impacts, responses, and resilience

Sbusiso Mbuyane, Portia Pearl Siyanda Sifolo and Rosa Naudé-Potgieter

8.1 Introduction

Travel agencies navigate through a world prone to disruptions, from pandemics, natural disasters, changing customer trends to political unrest. These disruptions have a significant economic impact, straining operations, and client relationships; hence, travel agencies need effective responses. By prioritizing communication, offering flexible solutions, and building trust, travel agencies may go beyond just navigating disruptions effectively and emerge even stronger by always being prepared for the unexpected and being equipped with existing business and travel trends. This further presents the complexity of travel disruptions that travel agencies face during any crisis, which has an impact on travel agency operations. The existing travel agencies have crisis management plans to ensure smooth communication and rebooking processes, while technology facilitates real-time updates and improved client support. However, the long-term resilience strategies, which include diversification of products and services offered and staff training in handling disruptions, are not heavily explored, and they will be looked at in this chapter.

8.2 COVID-19 pandemic and the travel agency industry landscape

COVID-19 negatively affected travel agencies worldwide including OTAs. Chen and Mansfeld (2021) mentioned that the largest seven public OTAs with an estimated market capitalization of US\$222.114 billion, such as Booking. com, Airbnb, Trip.com, Expedia, Makemytrip, webjet, and trivago, lost at least US\$11.5 billion in revenue in 2020/21 due to the virus (Dube, 2021). The lockdowns and restrictions forced office closures and a shift towards online bookings, disrupting operations and contributing to an estimated over 100 million job losses in the travel and tourism industry (Kusumaningrum & Wachyuni, 2020). With limited physical consultations, online platforms witnessed a surge, potentially impacting the traditional agency model of OTAs (Raza et al., 2021). However, after the world cautiously reopened, travel agencies adapted by focusing on domestic travel, offering flexible booking options, and

DOI: 10.4324/9781032720555-9

leveraging technology for online consultations and support (Rogerson, 2021). The path to recovery remained long, but travel agencies embraced digital transformation, adapted to changing consumer behaviour and prioritized flexibility to better navigate the new travel landscape.

8.3 Strategies and measures used to curb disruptions globally

Historically the travel and tourism industry has been characterized by the influence of external crises (Dube, 2021). Every crisis is a learning opportunity for the travel and tourism industry (Coombs, 2004). Every crisis presents the travel and tourism industry with a platform to integrate, learn, and improve their response strategies (Dionne, 2013). Table 8.1 outlines a brief overview of major crisis events and their area of origin that collectively influenced the performance of the travel and tourism industry and strategies adopted by governments to mitigate the crisis.

Strategies and measures used to deal with the external environment remain vital for the operation of the travel and tourism industry as they become an immediate reference point for future crisis events. Strategies and measures such as isolation, social distancing, and mass screening helped the industry to get into new capabilities to deal with more complicated situations in the future (Chen & Mansfeld, 2021). Hussain (2021) mentioned that unlike other rare financial crises, political crises, and civil wars, pandemics and epidemics occur regularly and repeatedly throughout history. With the outbreak of numerous health-related crises in the past and recent history, the travel and tourism industry worldwide has gained response experience, with the occurrence of SARS and H1N1, on how to embark on different recovery models and strategies as mentioned in Table 8.1 (Li et al., 2020).

8.4 Coordination between travel and tourism stakeholders

Coordination between travel and tourism stakeholders refers to the stakeholder engagement of the private and public sectors and the full cooperation between them for a comprehensive and cost-effective crisis recovery (Zhong et al., 2021). The travel and tourism industry is a complex network of millions of suppliers and consumers who trade experiences and services (Hussain, 2021). In looking at the resilience of the travel and tourism system, it is important to assess the level of collaboration, coordination, and synergy between industries (Richards, 2020). Matiza (2021) mentioned that the resilience of the travel and tourism industry requires a concerted multi-stakeholder approach to address both the travel and tourism industry demand and supply aspect. It is very common to see coordination between the industries, such as airlines negotiating new routes with National Tourism Organizations or hotels running marketing alliances with Local Destination Marketing Organizations (LDMOs) in an effort to increase visitation and build a more holistic destination image (Mizrachi & Gretzel, 2020).

Table 8.1 Medical/health/infectious crisis and their areas of origin

Name of crisis	Key stakeholders involved	Strategies adopted to manage it
Spanish flu H1N1	World Health Organization	Sanitary brigades
	United Nations Office for Project Services	National education campaign
		Established prescriptive regulations (1918)
Ebola	United National Children's fund	Social and physical distancing
	World Health Organization	Surface evacuation
		Surgical clothing (2014)
SARS-CoV1	World Health Organization	Isolation and quarantine
	Centre for Disease Control	Tracking
	Research and Prevention	Mass screening
	National Microbiology Laboratory	Hygiene (handwashing) (2002)
Middle East Respiratory Syndrome	World Health Organization	Tracking protocols
	National Institute of Allergy and	Rolling out of early testing
	Infectious Diseases, and Welcome Trust	Vitro diagnostic kits (2012)
SARS-CoV2	African Vaccine Regulatory Forum	Mass screening
	South East Asia Regulatory Network	Postponement of events
		Mass testing
	European Medicines Agency	Quarantine and isolation
	World Health Organization	Masks wearing
	Bill and Melinda Gates Foundation	Sanitizing (2020)

Source: Author's own compilation.

Travel agencies applied the proposed health protocols to facilitate safe travels and shifted focus to diversification of markets to avoid over-reliance on international tourism (Bama & Nyikana, 2021). For travel agencies to avoid immediate cash flow disturbances and to stimulate business and appeal to their customers, they had to modify the conditions for cancellations of travel bookings. Customers were now allowed to request a refund after 18 months for an un-used voucher (Mizrachi & Gretzel, 2020). Other practices that travel agencies adopted to revive their businesses according to Matiza and Slabbert, (2021) are as follows:

- Removal of the cancellation fees,
- Promoting flexible rates,
- Timely booking modifications,
- Switching their business model to conduct business on online platforms.

Li et al. (2020) mentioned that during and post a crisis, coordination between the travel and tourism industry generally improves and inspires great recovery. The element of communication, the resumption of marketing, and stakeholder cooperation are the key strategies that must be involved in the travel and tourism industry's recovery. Part of the coordination between travel and tourism stakeholders involves the travel and tourism industry's accumulated past crisis

Table 8.2 Travel agencies' traditional and modified strategies

Traditional strategies	*Modified strategies*
Dealing with crises independently	Forming new partnerships
Consultation	Online platforms and reduction of office consultation
Prioritizing fixed rates	Promoting flexible rates
Contractual cancellation fees	Removal of the cancellation fees

Source: Author's own compilation.

experiences for reference, involvement of inter-departmental efforts, and collaboration – tourism businesses need to come together and share information on current business trends – for a connected industry that is faced with the same problem needs to assist each other, and having a unified voice to represent the travel and tourism industry. To avoid a total business collapse, travel agencies had to modify their traditional business strategies to meet the demands of the crisis, and to build a future perspective to avoid possible complexities, the strategies are listed in Table 8.2.

The government's role as the biggest stakeholder in the travel and tourism industry in South Africa was to provide a platform for travel and tourism business stakeholders registered with the government (Rogerson, 2021). The platform was to build and maintain the capabilities of the travel and tourism industry to prevent, protect, mitigate, and respond to deep crisis damages and permanent impacts (Baubion, 2013; Rogerson, 2021). The direct intervention of the government was through the postponement of tax payments, assisting travel and tourism businesses with stimulus packages and later, temporary reductions in employer taxes to assist businesses in the travel and tourism industry to remain operational post the COVID-19 situation (Oberg, 2021).

8.5 Traveller messaging and maintaining confidence

Focusing mainly on the travel agencies, without looking at the client or the customer's response, would be a missed opportunity. Tourists are generally susceptible to cognitive biases, such as risk aversion when considering their tourism destination and product choices. Safety of tourism products and activities are examples of critical antecedents to tourist decision making and behaviour (Matiza & Slabbert, 2021). The relationship between pandemics and travel and tourism is central to understanding health security from both the customer and the travel agency's perspective and having a better understanding of global change-related issues such as weather patterns (Matiza & Slabbert, 2021). The COVID-19 pandemic presented a clear danger to tourists' health and well-being due to the virus being highly contagious (Matiza & Slabbert, 2021). In the era of COVID-19, the level of safety associated with tourism activities was imperative for both the travel and tourism industry recovery and stimulation of tourism demand throughout the travel and tourism value chain (Matiza & Slabbert, 2021).

The decline in South African traveller confidence was an indicator of how concerned tourists were about the state of the economy, which indicates that tourists considered mostly the theory of planned behaviour. The theory of planned behaviour assumes that individuals act rationally, according to their attitudes, subjective norms, and perceived behavioural control (Bama & Nyikana, 2021). To restore the traveller's confidence and stimulate demand, the government and all stakeholders within the travel and tourism industry also implemented new hygiene, safety procedures, and testing standards (Nyawo, 2020). Such measures reassured domestic travel safety and inspired the confidence of international travellers (Nyawo, 2020).

Prior to travel, tourists engage in meticulous research, consulting travel consultants or travel organizations to identify potential disruptions such as weather events, political unrest, or health concerns (Bama & Nyikana, 2021). The acquisition of travel insurance with comprehensive disruption coverage is highly recommended (Oberg, 2021). This security protects against unforeseen cancellations, delays, or medical emergencies (Rogerson, 2021). Tourists should prioritize flexibility by booking refundable or modifiable travel arrangements whenever possible (Dube, 2021). This adaptability allows for itinerary modifications in response to unforeseen disruptions. Additionally, downloading essential mobile applications and offline maps for the chosen destination proves invaluable (Nyawo, 2020).

Kapa et al. (2023) mentioned that expert training must be provided to tourism industry practitioners so that they can meet the professional standards and expectations of travel companies and tourists to overcome challenges that the industry faces. Training and education are critical to acquiring knowledge and skills such as interpretation and communication to perform a diversity of roles such as mediating between service providers and tourists (Hussain, 2021). By implementing these preparatory measures, tourists can approach travel disruptions with a sense of confidence and ultimately ensure a more secure and enjoyable travel experience. These measures will keep travel and tourism businesses up to date with the latest techniques for building long-term resilience (Kapa et al., 2023).

8.6 Responding to travel disruptions

Government agencies play a critical role when it comes to responding to travel disruptions or risk identification. In South Africa, the travel and tourism industry was further heightened by the emergence of COVID-19 variant N501Y on November 9, 2021 (Dube, 2021). The N501Y variant was dubbed the South African variant and caused many countries to impose a red flag alert for travellers from South Africa (Dube, 2021). Furthermore, the South African National Department of Tourism developed a recovery plan that proposed

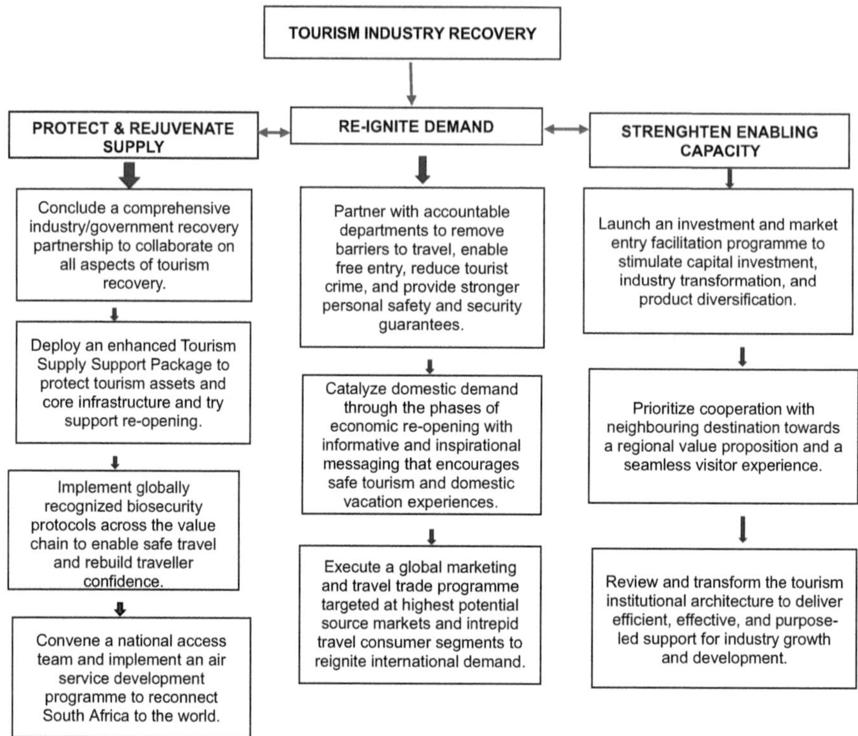

Figure 8.1 Department of Tourism Strategic recommendations.

Source: Department of Tourism, 2020.

three strategic themes or measures: rejuvenating supply, re-igniting demand, and strengthening enabling capacity (Nyawo, 2020). Figure 8.1 is a detailed plan that focuses strictly on how the travel and tourism industry amplified the supply and protected its core business during the crisis and existed after the crisis event, as recommended by the Department of Tourism. Strengthening capacity was to strengthen the industry beyond crisis value, ensuring future resilience and strengthening the travel and tourism industry value chain nationally.

The strategies adopted by different governments worldwide proved to be effective in restoring order, life, economy, and the travel and tourism industry back to normal (Oberg, 2021). When the negative effects of COVID-19 surged, governments instituted bans at the global, regional, and national levels, imposed travel restrictions, stay-at-home orders, mandatory quarantines, and other business restrictions at multiple levels of market and organizations (Raza et al., 2021). Inadvertently, South Africa, which had the highest number of infected people in Africa, entered into a five-stage lockdown regulated from stage one to five in a bid to save lives, and travel agencies had to operate after

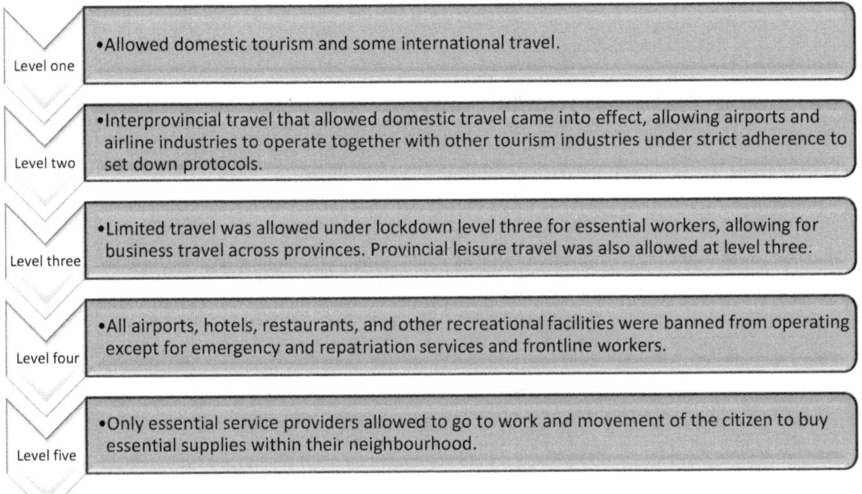

Figure 8.2 COVID-19 protocols.

Source: Adapted from the Department of Tourism (2020) and edited by the authors.

the crisis, with the adoption of technological updates in order to be competitive in the market (Dube, 2021). The lockdown levels were enforced on March 26, 2020. The levels of the lockdown were built to structure travel and tourism business operations, their return to full function, and to meet set standards at every stage of the crisis. The stages, ranging from one to five, are explained in Figure 8.2.

The protocols were enforced as per Figure 8.2 depending on the number of infections, the progress made regarding the containment of the virus, and the number of vaccinated people in the country (Matiza & Slabbert, 2021). The government protocols made a positive contribution to relaunching the travel and tourism industry, although the losses during the COVID-19 pandemic were still being felt (Bama & Nyikana, 2021). Non-adherence to treatment for diseases like HIV/AIDS and tuberculosis (TB) has had severe human, economic, and social costs to the government, hence South Africa reopened provincial and domestic travel to stimulate domestic tourists demand and indirectly restore international tourist confidence (Nyawo, 2020). From the above, it is evident that the travel agency industry is a key component of South Africa's growth strategy and will remain so in alleviating the travel and tourism industry.

8.7 Responses for general travel and travel agencies

Policies and travel restrictions implemented in several countries influenced the perception of tourists during and after the pandemic (Kusumaningrum & Wachyuni, 2020). Legal considerations changed rapidly depending on the

Table 8.3 Government policy response for the travel and tourism industry

Policies	Details
Socio-political	Vaccine passport requirements
	Traveller's placement in the red flag list
	Restrictions by other countries
Social	Retrenchments
	Tourists anxious to travel
	Stress of the travel and tourism business owners
Economic	Inability of travel businesses to service their debts
	Loss of income
	Salary cuts
	Ruined relationships with suppliers and customers

Source: Author's own compilation.

decisions taken by the government, and those decisions were changed following the level of complication of the pandemic (Kusumaningrum & Wachyuni, 2020). Legalities that were adopted in South Africa include vaccine/COVID-19 certificates, testing when entering another country, constant sanitizing, social distancing, and mandatory isolation if the traveller is from high-risk countries (Oberg, 2021). The legalities are further classified into three categories which are socio-political, social, and economic, and Table 8.3 displays the legalities accordingly.

The COVID-19 pandemic added to the already existing internal and external socio-political issues that the travel and tourism industry is faced with under normal circumstances (Godovykh et al., 2020). The government policies were a part of the recovery plan by the government for the travel and tourism industry, but due to the travel and tourism industry's heavy reliance on people's movements and interaction, the policies exposed the industry to even more complexities (Oberg, 2021). For resilience purposes, the South African small tourism businesses between 2008 and 2014 embarked on a six-year plan to deal with tourism-related issues (Dube, 2021). The plan included:

- Seeking government financial support, downsizing, and cost-cutting, approaching the domestic travel and tourism industry with inbound marketing,
- Encouraged solidarity-inclined travel and tourism, signing a new collective agreement between the industry and workers unions on pay-related matters,
- New public relations, shorter shifts, and work to sustain the industry,
- Developed an expansion strategy with a focus on crisis management, pursuing new and more resilient markets, establishing new marketing channels, and fostering government-private industry collaboration within the industry marketing forum.

Travel agencies face heightened legal and ethical considerations when formulating travel policies. Legally, they must adhere to updated regulations regarding travel restrictions, cancellation procedures, and potential health and safety protocols such as vaccine policies and visa regulations (Oberg, 2021). Ethically, their responsibility to the client well-being becomes even more crucial. This included offering transparent information about risks at destinations, implementing flexible booking policies to accommodate travel disruptions, and ensuring clients are aware of any health insurance requirements (Dube, 2021). Additionally, responsible travel practices gained greater importance: when responding to any disruption, travel agencies need to promote destinations with strong health and safety measures and prioritize tourism options that minimize environmental impacts (Zhong et al., 2021). By navigating this complex legal and ethical landscape, travel agencies can rebuild trust with clients, ensure their safety, and operate within a responsible framework in the travel environment.

8.7.1 Traveller behaviour and decision making

The main reason for the decrease in travel confidence is that people do not feel safe to travel (Hussain, 2021). As mentioned earlier in the literature, travellers consider planned behaviour – which deals with individuals acting rationally, according to their attitudes, subjective norms, and perceived behavioural control (Bama & Nyikana, 2021). A tourism crisis normally has a negative impact on the image of a tourist destination; in turn, that image lowers the intention of potential tourists to visit (Nyawo, 2020; Zhong et al., 2021). The travel and tourism industry has been innovative in winning people's confidence (Hussain, 2021). In this regard, innovative investment in the travel and tourism industry has become a crucial factor in giving people relevant information and multiple options to decide and motivate them to travel (Hussain, 2021). Travel and tourism industry businesses, such as travel agencies, offer packages with relevant information about safe travel and precautionary measures related to COVID-19, to boost their sales and decrease the level of panic from their clients (Hussain, 2021).

To facilitate travel, restore business and industry recovery travel agencies had to work with the transport industry, accommodation industry, and restaurant industry to promote high hygiene standards, combining the government recommendations and guidelines on social distancing, consistent sanitizing, and tracking of movements to restore customer trust (Hussain, 2021). Travel agencies had to monitor the impacts of the pandemic on the travel and tourism industry and the fast-evolving situation to stay alert and updated about new preventative measures, altered methods to conduct business, improved the number of people allowed into their premises, and new policy interventions from the government and industry players (Matiza & Slabbert, 2021).

8.7.2 Role of technology and social media in travel choices

Technology is the main tool for travel, starting from searching for information (preparation), during the trip and post the trip (Kusumaningrum & Wachyuni,

2020). New technology advances can empower the travel and tourism industry and bring new opportunities (Lu et al., 2022). In any adversity, there comes a period of innovation and business creativity. Crises may be short or may last for a long time, but when a tourism business has the current technologies in place, it is considered ready for any adversity, enabling it to thrive hard and survive during difficult times (Matiza & Slabbert, 2021).

Technological advancements have improved global connectivity and visibility of destinations, and the travel and tourism industry has become an extractive economy that sells an experience (Hussain, 2021). The travel and tourism industry has adapted technology in its operations, from music concerts performed via Zoom to museums, theatres, and cinemas putting their content online. The event industry offers an option of hybrid meetings, where some people are in a venue while the rest of the participants are online (Richards, 2020; Hussain, 2021). Airline companies such as Qatar Airways and Emirates Airlines offer customers contactless booking, contactless check-ins, self-service, digital boarding cards, and enhanced cleaning onboard (Hussain, 2021). These steps are taken to market contactless and remote economy and promote social distancing in the travel and tourism industry to increase travel confidence and motivate people to travel (Hussain, 2021).

The new technology is used to market and promote travel and tourism products to potential visitors without being physically there, which influences the visitor decision-making process in choosing to travel and has a significant contribution towards behavioural intention after the travel decision has been made (Lu et al., 2022). New technology and virtual reality can be used as a standalone entertainment tourism product, especially for theme parks, which have been applied in various theme parks worldwide, for example, Star Wars & Secrets of the Empire at Disney, VR Park in Dubai, VR park Star Park in Guizhou, China (Lu et al., 2022). The influence of the new technology and social media presence improves visitor experience, increases the level of interaction, and stimulates learning and comprehension (Lu et al., 2022). This helps tourism businesses to generate revenue without contributing to climate change, pandemics, and social unrest. New technology is also helpful for people with disability to enhance their travel and tourism experience through virtual accessibility (Lu et al., 2022). According to Mizrachi and Gretzel (2020) sales-focused marketing is less relevant now, as there is little demand to capture; instead, people need to know if it is safe to travel, and technology can help to mediate that.

8.8 Summary

This chapter has highlighted how travel agencies need the support of all their stakeholders during a crisis period. They need all the support they can get to recover, re-launch, and re-establish the travel and tourism value chain successfully with all the players involved and updated with the necessary business trends. The travel and tourism industry must be always ready with a reset button, should a crisis of any magnitude hit. To rethink and reposition itself with the new and emerging technologies, attracting new customers, and offering better services using strategies gained through various business phases to combat the potential problems. In the information and digital age, tourist's perception is strongly influenced by media reports and exaggerated by inaccurate information on the internet, which causes confusion; therefore, post the crisis recovery and measures, travel agencies must still work closely with media agencies to ensure that the travel and tourism industry communicate authentic and non-sensational reports. The travel and tourism industry, particularly travel agencies, has the responsibility to claim their importance within the travel and tourism industry space as the level of technology is increasing rapidly and presents itself as the option for the consumer. The digital era represents an era of new possibilities; therefore, the travel agency industry has the responsibility to revive the travel and tourism industry as they are the engine room for both the service and delivery of the industry.

8.8.1 Limitations

The literature on travel disruptions in travel agencies, as well as the resilient strategies in terms of the impacts they suffered during the COVID-19 pandemic, is not well documented among the developing countries. This chapter relies on South African examples. The chapter will contribute to the travel agency industry, which forms part of the travel and tourism industry. The chapter only focuses on tourist's perceptions regarding crisis events, and how tourists view tourists' destinations and not on the tourists' experience before, during, and after the pandemic. The level of digital involvement in the travel and tourism industry is minimally explored, which is the base and the future of the travel and tourism industry.

8.9 Conclusions

The chapter recommends that further studies probe the future of travel agencies' competitiveness in the digital age, the rate at which technology is advancing, and whether technologies such as AI and metaverse does limit or expand the functions of travel agencies. Incidents like SARS-CoV2 and the Ebola pandemic influenced the preferences and choices of tourists massively; therefore, future studies focusing on travel trends versus generational gaps and motivations behind them are critical to navigate the travel disruptions while developing resilient strategies for office-based and OTAs. Building resilience for travel agencies must involve government as their stakeholder for both policy and financial security. Existing measures to prevent crisis are in silos, and therefore, a comprehensive crisis management plan that involves all the industry players and the public sector will prove more resilience. This approach will also play a vital role in influencing the theory of planned behaviour. Developing countries like South Africa could benefit more from improving digital collaboration and inter-departmental cooperation, as this will help to shape the future of the industry during and after crisis events.

References

Bama, H. K. N., & Nyikana, S. (2021). The effects of COVID-19 on future domestic travel intentions in South Africa: A stakeholder perspective. *African Journal of Hospitality, Tourism and Leisure*, 10(1), 179–193. https://doi.org/10.46222/ajhtl. 19770720-94

Baubion, C. (2013). OECD risk management: Strategic crisis management. *OECD Working Papers on Public Governance*, 3, 1–24.

Chen, O. B., & Mansfeld, Y. (2021). A wasted invitation to innovate. Creativity and innovation in tourism crisis management: A QC&IM approach. *Journal of Hospitality and Tourism Management*, 46(20), 272–283. https://doi.org/10.1016/j.jhtm.2021.01.003

Coombs, W. T. (2004). *Research design: Qualitative, quantitative, and Mixed methods approach*. Los Angeles: SAGE Publication.

Department of Tourism (2020). Tourism industry survey of South Africa: COVID-19: Impact, mitigation and the future. Available at: https://www.tourism.gov.za/currentProjects/TourismrelieffundforSMMEs/Documents/Tourism%20Survey%20of%South%20Africa%20COVID-19.pdf [Retrieved 25 Jan 2024].

Dionne, G. (2013). Risk management: History, definition, and critique. *Risk Management and Insurance Review*, 16(2), 147–166. https://doi.org/10.1111/rmir. 12016

Dube, K. (2021). Implications of COVID-19 induced lockdown on the South African tourism industry and prospects for recovery. *African Journal of Hospitality, Tourism and Leisure*, 10(1), 270–287. https://doi.org/10.46222/ajhtl.19770720-99

Godovykh, M., Pizam, A., & Bahia, F. (2020). Antecedents and outcomes of health risk perceptions in tourism, following the COVID-19 pandemic. *Tourism Review*, 76(4), 737–748. https://doi.org/10.1108/TR-06-2020-0257

Hussain, A. (2021). A future of tourism industry: Conscious travel, destination recovery and regenerative tourism. *Journal of Sustainability and Resilience*, 1(1), 1–11.

Kapa, M. G., de Crom, E. P., & Hermann, U. P. (2023). Perceived challenges facing tourist guides in South Africa. *African Journal of Hospitality, Tourism and Leisure*, 12(2), 491–503. https://doi.org/10.46222/ajhtl.19770.381

Kusumaningrum, D. A., & Wachyuni, S. S. (2020). The shifting trends in travelling after the COVID-19 pandemic. *International Journal of Tourism & Hospitality Review*, 7(2), 31–40. https://doi.org/10.18510/ijthr.2020.724

Li, Z., Zhang, X., Yang, R. S., & Ciu, R. (2020). Urban and rural tourism under COVID-19 in China: Research on the recovery measures and tourism development. *Tourism Review*, 76(4), 718–736. https://doi.org/10.1108/TR-08-2020-0357

Lu, J., Xiao, X., Xu, Z., Wang, C. W., Zhang, M., & Zhou, Y. (2022). The potential of virtual tourism in the recovery of tourism industry during the COVID-19 pandemic. *Current Issues in Tourism*, 25(3), 441–457. https://doi.org/10.1080/13683500.2021. 1959526

Matiza, T. (2021). How to revive African tourism after the COVID-19 pandemic, viewed 27 January 2024, from https://www.google.com/amp/how-to-revive-african-tourism-after-the-COVID-19-pandemic-163832

Matiza, T., & Slabbert, E. (2021). Tourism is too dangerous! Perceived risk and the subjective safety of tourism activity in the era of COVID-19. *GeoJournal of Tourism and Geosites*, 36(2), 580–588. https://doi.org/10.30892/gtg.362spl04-686

Mizrachi, I., & Gretzel, U. (2020). Collaborating against COVID-19: Bridging travel and travel tech. *Information Technology & Tourism*, 2(5), 489–496. https://doi.org/10.1007/s40558-020-00192-0

Nyawo, J. C. (2020). Evaluation of government responses and measures on COVID-19 in the tourism sector: A case of tour guides in South Africa. *African Journal of Hospitality, Tourism and Leisure*, 9(5), 1144–1160. https://doi.org/10.46222/ajhtl. 19770720-74

Oberg, C. (2021). Conflicting logics for crisis management in tourism. *Journal of Tourism Futures*, 7(3), 311–321. https://doi.org/10.1108/jtf-10-2020-0191

Raza, M., Hamid, A. B. A. & Cavaliere, L. P. L. (2021). The e-tourism beyond COVID-19: A call for technological transformation. *Journal of Liberty and International Affairs*, 7(3), 118–139. https://doi.org/10.47305/JLIA2137118r

Richards, G. (2020). *Tourism in challenging times: Resilience or creativity.* Netherlands: Breda University of Applied Sciences.

Rogerson, J. M. (2021). Tourism business responses to South Africa's COVID-19 pandemic emergency. *GeoJournal of Tourism and Geosites*, 35(2), 338–347. https://doi.org/10.30892/gtg.35211-657

Zhong, L., Sun, S., Law, R., & Li, X. (2021). Tourism crisis management: Evidence from COVID-19. *Current Issues in Tourism*, 24(19), 2671–2682. https://doi.org/10.1080/13683500.2021.1901866

9 Understanding resilience concepts and frameworks for the tourism sector

Vanda Maráková, Michał Żemła and Genka Rafailova

9.1 Introduction

On a global scale, tourism is considered one of the most dynamic sectors in the world. Its development can significantly influence the economy of individual countries and, at the same time, create an opportunity for a more competitive position of the country as a tourism destination. Recently, however, the tourism sector was/is facing various global crises (COVID-19, energy crisis, war conflicts, climate changes, etc.), which reveal the intensity of its vulnerability and, at the same time, test its resilience. Resilience is the subject of academic discussion (Neise et al., 2021; Nöhammer et al., 2022; Ntounis, et al., 2022; Yang et al., 2023, etc.) as well as the concern of international organisations (OECD, 2022; WTTC, 2022) and the agenda of policymakers.

The resilience of tourism can be defined as the ability of the social, economic, and ecological systems of individual countries to recover from multiple shocks and stresses resulting from past and current crises, which tend to negatively affect the development of tourism (Yang et al., 2023). Other interpretations (Neise et al., 2021; Nöhammer et al., 2022; Ntounis, et al., 2022) of tourism resilience emphasise the ability of destinations not only to quickly recover from shocks but also to anticipate these shocks and implement effective measures and strategies, which can either prevent them or mitigate the impact of negative impacts on the tourism sector.

There are differences in the pace of tourism recovery between countries and regions. Differences in the speed of recovery and restoration of the tourism sector depend on the tourism policies of individual countries and on the amount of government expenditures serving in the restoration of the tourism sector at the national level (Yang et al., 2023). Activities stimulating the pace of recovery of the tourism sector are often part of tourism development strategies, which are formed either by national tourism organisations but also by regional and local tourism organisations.

According to UNWTO (2022) and WTTC (2022), the achievement of economic resilience in the tourism sector is conditional on the sustainable growth of tourism consumption. Hall (2013) and Sharpley (2021) further refer to the fact that the growth in domestic and foreign visitor spending should not be

DOI: 10.4324/9781032720555-10

driven by the growth in the visitor volume but by other incentives that will ensure a change in consumer behaviour.

The resilience of the sector was investigated at all levels, while the studies, documents, and strategies themselves were most often focused on the resilience of tourism at the national level. However, less attention was paid to the investigation of the vulnerability of tourism at the regional and local levels. At the local level, the authors (Neise et al., 2021; Ntounis et al., 2022; Yang et al., 2023) investigated the resilience of tourism businesses, while at the regional level (Nöhammer et al., 2022; Liu et al., 2023; CMTS 2020) and national level (Hailemariam & Dzhumashev, 2023; UNWTO, 2022; OECD, 2022; WTTC, 2022; IBRD, 2020; AFMST, 2019; SMITT, 2019) studies focus on a mezzo and macro-approach.

9.1.1 Tourism resilience in selected countries

Austria, Malta, Croatia, and Spain are indicated to be the countries with the highest average expenditure of visitors per stay (OECD, 2022). Each of these countries implemented during the pandemic the resilience policies, which are analysed below. Several activities result from the development strategies of selected EU countries (Austria, Malta, Croatia, Spain), which purposefully influence the growth of consumption in the tourism industry, thereby ensuring a lower level of economic vulnerability in the tourism industry. The activities of the given countries (Table 9.1) are primarily aimed at ensuring sustainable consumption growth in tourism, which contributes to achieving economic resilience in the tourism sector. By considering this, we decided to focus on these four selected European countries as they can serve as a benchmark for other countries that are lacking in introducing policy instruments to address the tourism resilience strategies. We have conducted the content analysis of relevant strategic documents.

The Austrian Federal Ministry of Sustainability and Tourism (AFMST, 2019) in PLAN T lists several activities aimed at achieving sustainable and resilient tourism development in the destination Austria, but only a few of them serve primarily to stimulate the growth of tourism consumption. Providing higher wages, career growth, and adequate training processes to employees in tourism establishments contributes to the attractiveness of work in the tourism sector in Austria (AFMST, 2019), Croatia (CMTS, 2020), and Malta (MTM, 2021). The importance of increasing the attractiveness of work in the tourism sector in examined countries lies mainly in the growth of employment in the tourism industry, which is subsequently reflected in the growth of consumption in the tourism industry.

The Malta Ministry of Tourism (MTM, 2021) developed a tourism development strategy that defined several activities aimed at ensuring the stimulation of visitor spending in domestic and inbound tourism, through which they want to contribute to a more resilient tourism sector. One of the country's priority tasks is to ensure an increase in government spending (e.g. subsidies),

Table 9.1 Activities ensuring the economic resilience of tourism in countries with the highest average expenditure of visitors per stay

Strategy	Activities ensuring the economic resilience of tourism
Austria (AFMST, 2019)	– the attractiveness of work in the tourism sector; – strengthening of cooperation between typical tourism enterprises and enterprises of other sectors (agriculture, food industry, etc.).
Malta (MTM, 2021)	– reduction of seasonality; – the attractiveness of work in the tourism sector; – investments in the development of tourism at the regional and local level; – growth rate of average expenditure per visitor > growth rate of visitors;
Croatia (CMTS, 2020)	– reduction of seasonality; – the attractiveness of work in the tourism sector; – investments in the development of tourism at the national level; – growth in the provision of accommodation services operating on the principle of a shared economy and the possibility of camping.
Spain (SMITT, 2019)	– sustainable growth of tourism at the national level; – increasing the prosperity of tourism facilities at the regional and local levels; – growth rate of average expenditure per visitor > growth rate of visitors.

Source: Own elaboration, 2024.

which will serve as investments in the innovation and modernisation of existing tourism products and businesses located in destinations with a low frequency of foreign visitors. Investing in the innovation of new or existing tourism products in less visited destinations is intended to ensure a more even dispersion of visitors in inbound tourism, thus preventing mass tourism in the most visited destinations in the country while increasing tourism consumption in less visited destinations. The support and development of cultural tourism and organised events also play a key role in achieving a positive consumption stimulus in internal tourism in Malta. The development of cultural tourism and organised events should ensure an increase in the share of short-term stays of foreign visitors in Malta, even outside the peak season, thereby achieving a reduction in seasonality and an increase in the volume of consumption in the tourism industry. The development of cultural tourism and organised events in Malta should, among other things, contribute to the development of domestic tourism. According to the Malta Ministry of Tourism (MTM, 2021), it is also important to ensure a situation where the average expenditure per visitor grows faster than the number of visitors to the country, making the country more sustainable and resilient to crises.

The Croatian Ministry of Tourism and Sports (CMTS, 2020) developed a tourism development strategy with similar activities stimulating tourism consumption. Attention was focused mainly on the increase in government spending, which should be used to modernise all accommodation facilities in Croatia with the aim of increasing the quality of the services provided. In the regional action plan, the reference has been made to the reduction of seasonality as in Malta. However, they adapt the product of cultural tourism as a priority to the needs of domestic visitors in order to ensure the development of domestic tourism, which will explicitly affect the growth of consumption in domestic tourism.

The Ministry of Industry, Trade, and Tourism (SMITT, 2019) in Spain has focused its strategic plan on the sustainable growth of tourism in the country as a priority. The main goal of the country is to maintain the increasing tendency of the expenditure curve of visitors in domestic tourism, even under the conditions of a decreasing number of visitors in the destinations that are the most visited in the country. To maintain a balance between the three fundamental pillars (economic, social, and environmental), they turned their attention to financing destinations with low tourism consumption. The financing is intended to serve as an investment in the innovation and modernisation of existing tourism products and businesses located in destinations with a low frequency of foreign visitors. The goal is a more even distribution of foreign and domestic visitors in the country and an increase in the prosperity of tourism facilities in visited destinations, which will contribute to increasing the well-being of residents and economic growth at the regional and local levels.

9.1.2 Promoting resilient tourism destinations

OECD's (2022) study stresses the importance of diversification of regional economies. Destinations, as well as tourism businesses, need to ensure that the product has broad appeal to be able to generate a critical mass of visitors. This includes the spatial spread of tourists and attractions. Diversification can involve adapting the tourism product to reduce seasonality.

Strong destination management is an essential tool to stimulate destination resilience (Traskevich & Fontanari, 2023). There is an urgent need for a transition away from destination marketing towards destination management and stewardship (Reinhold et al., 2023). Identification of core competencies and key capacities (Prahalad & Hamel, 2009) that need to be strengthened is an emerging task. Inevitable is to include diverse stakeholders and thus develop more holistic business models (Freudenreich et al., 2020). Community, environmental, and economic aspects of resilience should be recognised and defined as the long-term benefits (Reinhold et al., 2023). The framework of long-term support should be developed to secure investments for resilience programmes and projects.

International organisations like OECD (2022) and WTTC (2022) underline that there is an overlap in policies that make economies resilient and destinations prosperous in terms of developing policies, measures, and instruments to support a dynamic business environment, diversified industrial base, and local skills development. Resilient destinations also promote sustainable development, well-being, and inclusive growth. Gupta and Vegelin (2016) and Hall et al. (2017) point out the importance of the destination level in ensuring resilience since it is a space where all different sub-sectors and actors come together. In a conscious destination, parties strive for an outcome that has the most positive possible social, ecological, and economic impact and leads to the highest possible quality of life, work, and experience.

As far as general, universal guidelines for destinations implementing resilience thinking in their approach to tourism management are being developed and are available for managers, the influence of specific features of destinations is not approached yet in the scientific debate.

9.2 Tourism resilience in the policy of three selected destination

9.2.1 Research methodology

The case studies in the article present tourism destinations in three European countries. The methodological approach for selecting the destinations is backgrounded in the research orientation to contribute to the tourism resilience study with recommendations for strategies or policies for developing tourism resilience. Case studies aim to disclose reactions to crises, achievements, and actions needed to build a resilient tourism ecosystem in destinations with certain similarities (belongingness to the political systems before 1990, economic transformation during political reforms after 1990, and the significance of industry as a driving force for urban development) and distinctiveness (geographical location, history, approach, type, and scale of tourism development). Following the listed criteria, three urban destinations in countries from the former socialist system were selected. Table 9.2 overviews the selected destinations: Košice (Slovakia), Katowice (Poland), and Varna (Bulgaria). They are similar in terms of surface area and number of inhabitants but different in terms of scale (number of hotel facilities and beds) and pathway of tourism development (seasonality and concentration of tourism supply). They are in Central and Eastern Europe, near the mountains (Košice), central (Katowice) and coastal (Varna) regions. The destinations present different types of tourism – seasonal and recreational (Varna), business and event (Katowice), and urban and cultural (Košice). Katowice and Košice are destinations with a shorter tourism history, and Varna is a well-established destination in the European tourism market, with a developed tourist infrastructure and a considerable contribution of the sector to the economy. In all three cities, the economic transformation has been caused by the decline of key sectors after 1990.

In the three destinations, in-depth interviews were conducted with consultants and heads of Destination Management Organisations – Visit Košice, Katowice Municipality, and the Tourism Department in Varna Municipality. The respondent from Košice is the executive director of Visit Košice DMO and chairwoman of the Association of Tourism Organisation in Slovakia. As for

Table 9.2 Comparison of the tourism destinations Košice, Katowice, and Varna by tourism, urban, and demographic statistics for 2023

Numbers	Košice	Katowice	Varna
Surface	243.7 km²	164.73 km²	237.5 km²
Inhabitants	239 141	280 190	341 516
Density – persons/km²	1008.9	1709.1	1437.9
Hotel facilities	138	24	270
Hotel beds	7,464	2,650	55,640
Tourists	218 973	426 728	1 071 428
Ratio of foreign tourists	0,43	0,27	0,70

Source: Own elaboration, 2024.

the Katowice, it was the author of the city's tourism development strategies. In Varna, we approached the Deputy Mayor, who manages the Tourism Department in Varna Municipality.

During the interviews, two main topics were discussed: (1) existing resilience strategies or plans and (2) preparation for potential disruptions or crises (e.g. risk assessments and contingency planning). The results were compared in relation to the distinctiveness of destinations and achievements in the European countries with the best average spending per visit.

9.2.2 Košice, Slovakia

Košice is the second biggest city in Slovakia, situated in the eastern part of the country.

The city has a strong economic background and a tradition of steel production. Besides steel, the IT sector is growing, as evidenced by the IT Valley cluster.

As of the current assessment, Košice has begun integrating resilience strategies within its broader tourism and city planning frameworks, though a formalised, standalone resilience plan specifically for tourism is not yet in place. The city's approach to resilience is primarily embedded within its disaster management and sustainable development policies. Key components of this integrated strategy include enhancing infrastructure resilience to ensure continuity of services during crises, promoting sustainable tourism practices that contribute to economic and environmental sustainability, and implementing comprehensive crisis management protocols to swiftly respond to and recover from various disruptions. Additionally, Košice collaborates with national and international bodies to align its resilience practices with broader standards and to gain insights into effective resilience building.

Košice employs a multi-faceted approach to assess and prepare for potential disruptions or crises, focusing on both pre-emptive risk assessments and contingency planning. This approach begins with regular risk assessments conducted in partnership with local government bodies, tourism operators, and emergency services. These assessments identify potential vulnerabilities and threats specific to the region, such as natural disasters, economic fluctuations, or health-related emergencies.

Based on these assessments, Košice develops contingency plans tailored to address identified risks. These plans encompass a range of responses from immediate crisis management to longer-term recovery strategies. They include establishing communication protocols to quickly disseminate information to businesses, tourists, and residents, creating backup systems for critical infrastructure, and training key personnel in crisis management and response.

Furthermore, Košice regularly revises and updates its crisis management plans to adapt to new information or changing circumstances, ensuring that the destination can respond effectively to unexpected situations. Workshops and simulations are conducted periodically to prepare and train stakeholders, enhancing the overall resilience of the tourism sector.

9.2.3 *Katowice, Poland*

Katowice is a city located in the southern part of Poland. It's the heart and the biggest city of the largest conurbation (Upper Silesian Conurbation) in Poland, which has about 2 million citizens (Szubert et al., 2021). Most of the cities in the conurbation, including Katowice, are young industrial cities. After the fall of communism and the heavy industry collapsed in the region, Katowice walked through the process of transitioning smoothly. Nowadays, it is a modern city with well-developed administrative, academic, and hi-tech functions. However, it is still experiencing image problems, and industrial stereotypes limit its attractiveness to domestic tourists (Adamus-Matuszyńska et al., 2019; Szubert & Żemła, 2019). Typical leisure urban tourism almost does not exist in the city (Szubert et al., 2021). Instead, it is the country's strongest centre of business tourism (Cieślikowski, 2024). After 2000, the city developed as a business tourism and events city with a unique infrastructure for conferences and congresses.

As a result, contemporary tourism in Katowice is shaped in a specific way and requires a specific attitude from the policymakers. The city has a limited number of hotels, but they are rather big, and the four-star hotels prevail (Cieślikowski, 2024). These facts lead to a situation in which local tourism planning involves a very low number of stakeholders, which fosters cooperation. Additionally, it must be noted that tourism does not play a key role in the city's economy; however, it is perceived as an important direction of diversification of the local economy and an opportunity to refresh the city's image.

The word 'resilience' does not exist in any official document regarding tourism in Katowice. The idea of resilience in tourism is also not familiar to city decision-makers. However, one can find few manifestations of resilience thinking. This is mainly the continuous search for diversification of tourism in the city. After successfully implementing sporting and music events, Katowice is developing a city break offer to minimise the high dependency on contemporary markets. This offer is expected to bring more domestic visitors to the city than big congresses.

Another absent element is preparation for future crises and shocks. Tourism in Katowice still plays a minor role, and crises such as COVID and war in the Ukraine are perceived as a threat mainly to other economic activities. The city's economy has transformed from heavy industry to high-tech industries and other modern production sectors, with the introduction of tourism-related economic activities perceived as a diversification of the economy and, in a way, a search for resilience in the local economy.

9.2.4 *Varna, Bulgaria*

Varna, the third largest city in Bulgaria, is a tourism destination on the Black Sea coast. The town has key tourist attractions such as the sea, beaches, mineral waters, and significant recreational nature resources. After the radical political reforms in 1990–2000, Varna's economy changed due to the decline of some marine industry sectors, which had an essential share in the city's GDP and employment. These processes and the strong dependence on seasonal seaside holiday tourism have prompted the Municipality's local government and Tourism Department to redirect the strategy and marketing efforts to develop year-round tourism by stimulating cultural, spa, dental, and sports tourism. As a result, in the last 5–6 years, spa and wellness tourism has developed throughout the year, but it is still concentrated in a few hotels in nearby resorts. Also, MICE tourism has been getting more intensive in the spring and fall.

According to the Tourism Department, Varna has no formal strategy devoted to tourism resilience. An effective strategic document of close importance is the Program for Sustainable Development of Tourism on the territory of the Municipality of Varna 2021–2030, which is annually updated. As an orientation to building the resilience of the destination, the Deputy Mayor considers the outlined priorities in the document:

1   Creating a safe and secure environment for tourism
2   Management and preservation of tourist resources
3   Development of the general and tourist infrastructure
4   Development of tourism in four seasons

These priorities fit well in the resilience framework, but their achievement depends on the city's infrastructure modernisation and integrative efforts of multiple sectors. The Plan for Integrative Development of the City Varna 2021–2027 (PID Varna) includes projects interconnected with work on the upbuilding of tourism infrastructure and preservation of tourism resources, considered in the Program for Sustainable Development of Tourism. Their focus on strengthening the coast, protecting it from sewage pollution, and preserving and socialising cultural heritage supports tourism resilience development.

The Consultative Council on Tourism (CCT), a stable public-private collaboration, works with representatives from the municipal administration and the tourism business and allows for a more in-depth consideration of the important tourism problems in Varna and decision-making on proposals to the Municipal Council.

Some of the reactions of the Municipality of Varna to crises are short-term oriented – reducing the fiscal burden, allowing exemption from fees, and other measures aimed at relieving the tourism industry. The policy for tourism development has refocused on domestic and nearby markets by redirecting advertising activities. The intention is to compensate for lost positions in issue markets while preventing the strong dependence on traditional foreign markets in the long term. Digitalisation is another step towards overcoming negative crisis consequences for tourism and guaranteeing its more stable future development.

An important action in the direction of assuring tourism resilience is accurate risk assessment, with regular examination of the following endogenous and exogenous factors, clarified during the interview:

1 Infrastructure problems.
2 Pathway of tourism development – the strong concentration of tourist supply and the pronounced seasonality.
3 The delay of implementation of the PID Varna 2021–2027 and of development of cultural spaces, construction of new convention centres, and renovation of museums and archaeological sites.
4 Geopolitical situation.
5 Climate change – hot summers, unpredictable storms.
6 Lack of qualified labour (migration of young people, demographic challenge).

In 2024, following the recommendation of CCT, a Public-Private Fund for the Support and Expansion of Varna Airport was established for direct support of flight programmes for the extension of the season.

The positive step in the direction of developing tourism resilience and sustainability is an evolving vision of the stakeholders for the future development of Varna's destination. It is based not only on increasing the volume of arrivals and spent nights but also on integrating tourism into the development of the local blue economy, as well as the creative, cultural, and IT industries.

9.3 Discussion and conclusions

To identify the activities that ensure the resilience of tourism at the national level, we analysed the available tourism development strategies of the EU countries (Austria, Malta, Croatia, and Spain), which during the first year of the COVID-19 pandemic had the highest average expenditure of visitors per stay compared to other EU member countries (OECD, 2022). The development strategies of selected EU countries resulted in several activities presented in Table 9.1, which purposefully influence the growth of consumption in the tourism industry, thereby ensuring a lower level of economic vulnerability of the tourism industry. As for the destination level, we followed the agenda of international organisations, mainly OECD (2022) and WTTC (2022), that have defined the framework of managing the tourist destination for a more resilient, sustainable, and inclusive future. The key policy considerations include reconsidering perceptions of tourism success, adopting an integrated policy-industry-community approach, mainstreaming sustainable policies and practices, developing more sustainable tourism business models, and implementing better measurements for better management (OECD, 2022).

Presented examples of cities with similar historical, social, and economic backgrounds, however, with very different tourism development patterns prove that resilience is to be perceived as an adaptive paradigm (Hunter, 1997). Different cities, with different levels of tourism development and different forms of tourism dominating, perceive the role of tourism in their economies differently and, as a result, define sustainable development of tourism in distinct ways. Košice, as a historical city, searches for the balance between the needs of residents and visitors and the tourism business and for integrating tourism with the development of the city. In Katowice, tourism is treated not as a crucial part of the local economy but rather as a tool for modernising this economy and refreshing the image of the city. Oppositely, in Varna, a city strongly dependent on mass recreational tourism, the key point in the sustainable future is natural assets protection. These differences are reflected in the attitude towards resilience shown by local authorities in the cities. Systemic resilience thinking is adopted by the Košice authorities. However, the city is at the beginning of the process of implementation of this. In Varna, authorities strive for just several elements of a resilience system. There is a need to construct and develop proper infrastructure to achieve better economic and ecological resilience, which is key for coastal destinations. For example, Varna shares the sea with other countries and destinations, which means that any problem in one destination (pollution, political conflict, etc.) is a risk for tourism in others. Additionally, Varna is a very vulnerable destination because of seasonality and high dependence on tourism. However, considering these elements and risk preparations regarding them are rather visible symptoms than implemented systemic resilience thinking as there is no proactive and integrated governance of tourism. Finally, Katowice tries to develop tourism in a way that boosts the local economy the best, and the resilience thinking lies mainly in securing the tourism input into local development rather than tourism planning itself. This appointment to the adaptiveness of the resilience paradigm is the main output from the conducted comparison, as in the contemporary literature (OECD, 2022; Yang et al., 2023) the search for the universal theory dominates. This approach also brings significant hints for policymakers who not only should focus on finding a benchmark solution to implement in their destination but also should understand properly the specific features of this destination and their consequences for the resilience policy.

The chapter researches the differences in resilience policy between different cities. It is based on the analysis of three purposely selected examples, and more research in the future is necessary to establish a theory explaining in detail these differences. The resilience of subsystems determines the resilience of the whole system. Therefore, complexity and systems thinking should be applied to develop tourism resilience.

Acknowledgements

This work was supported by the Scientific Grant Agency of the Ministry of Education, Science, Research and Sport of the Slovak Republic VEGA under Grant number 1/0360/23 'Tourism of the New Generation – responsible and competitive development of Tourism destinations in Slovakia in the post-COVID era.'

References

Adamus-Matuszyńska, A., Michnik, J., & Polok, G. (2019). A systemic approach to city image building. The case of Katowice city. *Sustainability*, *11*(16), 4470. https://doi.org/10.3390/su11164470

Austrian Federal Ministry of Sustainability and Tourism. (2019). [online]. [cit. 2024.03.15]. Accessible on the internet: www.bmnt.gv.at/masterplan-tourismus

Cieślikowski, K. (2024). Business tourism market in Katowice. Report 2023. Business Destination Alliance, Katowice.

Croatian Ministry of Tourism and Sport. (2020). [online]. [cit. 2024.03.16]. Accessible on the internet: https://mint.gov.hr/UserDocsImages/arhiva/Strategy-tourism.present.pdf

Freudenreich, B., Lüdeke-Freund, F., & Schaltegger, S. (2020). A stakeholder theory perspective on business models: Value creation for sustainability. *Journal of Business Ethics*, *166*(1), 3–18. https://doi.org/10.1007/s10551-019-04112-z

Gupta, J., & Vegelin, C. (2016). Sustainable development goals and inclusive development. *International environmental agreements: Politics, Law and Economics*, *16*, 433–448. https://doi.org/10.1007/s10784-016-9323-z

Hailemariam, A., & Dzhumashev, R. (2023). The impact of pandemic-induced uncertainty shock on tourism demand. *Current Issues in Tourism*, *26*(16), 2575–2581. https://doi.org/10.1080/13683500.2022.2113044

Hall, C. M., Prayag, G., & Amore, A. (2017). *Tourism and resilience: Individual, organisational and destination perspectives*. Channel View Publications. https://doi.org/10.21832/9781845416317

Hall, M. (2013). Framing behavioral approaches to understanding and governing sustainable tourism consumption: Beyond neoliberalism, "nudging" and "green growth"? *Journal of Sustainable Tourism*, 21 (7), 1091–1109. https://doi.org/10.1080/09669582.2013.815764

Hunter, C. (1997). Sustainable tourism as an adaptive paradigm. *Annals of Tourism Research*, *24*(4), 850–867. https://doi.org/10.1016/S0160-7383(97)00036-4

IBRD. (2020). [online]. [cit. 2024.03.16]. Accessible on the internet: https://documents1.worldbank.org/curated/en/328421604042124972/pdf/Resilient-Tourism-Competitiveness-in-the-Face-of-Disasters.pdf

Liu, Y., Chen, L., Han, F., & Zhang, T. (2023). Regional tourism resilience under crisis impacts: The cases of Yangtze River Delta and Pearl River Delta. *Journal of Geographical Research*, *6*(4), 11–25. https://doi.org/10.30564/jgr.v6i4.5942

Ministry of Tourism, Malta. (2021). [online]. [cit. 2024.03.16]. Accessible on the internet: https://issuu.com/visitmalta/docs/maltatourismstrategy_2030_v7

Neise, T., Verfürth, P., & Franz, M. (2021). Rapid responding to the COVID-19 crisis: Assessing the resilience in the German restaurant and bar industry. *International Journal of Hospitality Management, 96,* 102960. https://doi.org/10.1016/j.ijhm.2021.102960

Nöhammer, E., Haid, M., Corradini, P., Attenbrunner, S., Heimerl, P., & Schorn, R. (2022). Contextual factors of resilient tourism destinations in a pandemic situation: Selected cases from North and South Tyrol during the SARS-CoV-2 pandemic. *Sustainability, 14*(21), 13820. https://doi.org/10.3390/su142113820

Ntounis, N., Parker, C., Skinner, H., Steadman, C., & Warnaby, G. (2022). Tourism and hospitality industry resilience during the Covid-19 pandemic: Evidence from England. *Current Issues in Tourism, 25*(1), 46–59. https://doi.org/10.1080/13683500.2021.1883556

OECD. (2022). OECD tourism trends and policies. Chapter 2. Building resilience in the tourism ecosystem. [online]. [cit. 2024.04.č]. Accessible on the internet: https://www.oecd-library.org/sites/efa87744-en/index.html?itemId=/content/component/efa87744-en

Prahalad, C. K., & Hamel, G. (2009). The core competence of the corporation. In M. H. Zack (Ed.), *Knowledge and strategy* (pp. 41–59). Routledge. https://doi.org/10.1016/B978-0-7506-7088-3.50006-1

Reinhold, S., Beritelli, P., & Laesser, C. (2023). The 2022 consensus on advances in destination management. *Journal of Destination Marketing & Management, 29,* 100797. https://doi.org/10.1016/j.jdmm.2023.100797

Sharpley, R. (2021). On the need for sustainable tourism consumption. *Sage Journals, 21*(1), 96–107. ISSN 2158-2440.

Spanish Ministry of Industry, Trade and Tourism. 2019. [online]. [cit. 2024.03.17]. Accessible on the internet: https://turismo.gob.es/es-es/estrategia-turismo-sostenible/paginas/index.aspx

Szubert, M., Warcholik, W., & Żemla, M. (2021). The influence of elements of cultural heritage on the image of destinations, using four Polish cities as an example. *Land, 10*(7), 671. https://doi.org/10.3390/land10070671

Szubert, M., & Żemła, M. (2019). The role of the geographical textbooks in grounding negative stereotypes of a tourism destination—The case of upper Silesian conurbation in Poland. *Administrative Sciences, 9*(2), 42. https://doi.org/10.3390/admsci9020042

Traskevich, A., & Fontanari, M. (2023). Tourism potentials in post-COVID19: The concept of destination resilience for advanced sustainable management in tourism. *Tourism Planning & Development, 20*(1), 12–36. https://doi.org/10.1080/21568316.2021.1894599

UNWTO. (2022). Impact assessment of the COVID-19 outbreak on international tourism. [online]. [cit. 2024.04.4]. Accessible on the internet: https://www.unwto.org/impact-assessment-of-the-covid-19-outbreak-on-international-tourism

WTTC. (2022). Enhancing resilience to drive sustainability in destinations [online]. [cit. 2024.03.16]. Accessible on internet https://wttc.org/Portals/0/Documents/Reports/2022/WTTCxICF-Enhancing_Resilience-Sustainable_Destinations.pdf?ver=2022-06-13-213556-557

Yang, S., Lu, Y., & Wang, S. (2023). Quantitative simulation and verification of the tourism economic resilience in urban agglomerations. *Scientific Reports, 13*(1), 18879. https://doi.org/10.1038/s41598-023-46166-0

10 Conceptualising multi-stakeholder resilience building

Shameena Fernando and Neil Carr

Introduction

Hazardous events that disrupt the functioning of society at any scale are referred to as disasters (UNDRR, 2024a). The COVID-19 pandemic and Turkey-Syria earthquake in 2023 are a couple of examples of natural disasters while the ongoing Russia-Ukraine and Israel-Palestine wars are examples of human-made disasters. These have all been identified as high-scale disasters in recent years (CRED, 2024). This label is associated with the number of casualties and injuries, the people affected, and the size of the financial damage associated with infrastructure and other assets (Noy, 2015). The importance of addressing disaster management in the context of tourism is related to the increasing number of disasters across the globe and the extent of their impact (UNDRR, 2022). Disasters disrupt continuity. Their effects and the duration, both of the event itself and its aftermath, are, at least partially, determined by the efforts put into disaster management phases.

Emergency management phases, widely known as the disaster management cycle (indicated in Figure 10.1), are used to investigate the causes of disasters and develop plans to reduce their impact (Mushkatel & Weschler, 1985).

i Mitigation

The impacts of natural and human-made disasters cannot be avoided fully. However, the scale and severity of the impacts can be minimised through disaster risk reduction. This aims to prevent new or existing disasters by establishing strategies and actions to identify and manage the causes of the disasters and their impacts before they occur. Mitigation measures comprise raising public awareness, environmental and social policy formation, development of capacity and infrastructure, and updating all types of civil engineering techniques (UNDRR, 2023b).

ii Preparedness

This phase focuses on the intelligence and capabilities developed by governments, response and recovery groups, and society to effectively predict, respond to and recover from the impacts of disasters that are likely to occur or have already occurred (UNDRR, 2023c). Actions carried out in

DOI: 10.4324/9781032720555-11

this stage are within the scope of risk management as governments and other relevant agencies aim to build and strengthen the capacity to effectively manage disasters (Mushkatel & Weschler, 1985). This stage includes developing and establishing contingency planning and logistical readiness. Gearing up with field training and arranging coordination, setting up early warning systems and providing information to the public are further actions often undertaken in this phase. Engaging the public in disaster drills is a part of preparedness strategies (UNDRR, 2023c).

iii   Response

This phase focuses on the immediate actions taken after a disaster (UNESCO, 2023). It includes rescue efforts, emergency healthcare, identification of casualties, and housing and supporting individuals who are impacted. During this phase public authorities and volunteer teams work to stabilise the situation minimising further harm (Asian Disaster Reduction Center, 2022).

iv   Recovery

This phase focuses on improving individual and community resilience post-disaster (Center For Disaster Philanthropy, 2023). It involves planning and developing policies to achieve sustainable development through future disaster risk reduction. This relates to the economy, environment, culture and society (UNDRR, 2023d).

Figure 10.1 Four Phases of Disaster Management Cycle.

Strategies that are implemented at each phase of disaster management have the potential to minimise the scale of the impact of a disaster. They also determine the pace of recovery from a disaster. These phases shape the tourism industry. Changes that are associated with disasters at each phase are felt by multi-stakeholders, such as policymakers (governments), businesses (directly and indirectly linked to tourism), tourism personnel, financial stakeholders and communities (both human and more-than-human). However, the prevailing policies associated with disasters and tourism are dominated by a single stakeholder perspective, that of governments (Das, 2023). Therefore, most existing policies fail to identify and address the challenges faced by multi-stakeholders across multi-levels from multiple perspectives. Instead, they force one perspective onto others in a top-down approach to governance (Azad et al., 2019). This is a serious gap that needs to be addressed for effective resilience building.

Resilience refers to the ability of a system. It is the ability of a society exposed to a disaster to resist, absorb, accommodate and recover from the impact of a disaster in a timely and efficient manner (UNISDR, 2012). Destination resilience building is vital for the tourism industry, as disasters and tourism are intimately connected. The latter impacts tourists' willingness to visit destinations (Rossello et al., 2020) and affects businesses to invest in these places (Smith & Carmichael, 2006). The COVID-19 pandemic is the latest large-scale disaster to halt the operations of the tourism industry. Post COVID-19 most countries and businesses encountered the challenge of building resilience. To achieve this, existing tourism disaster management policies should be developed in a way that fits all stakeholder needs. This need forms the basis for this conceptual chapter that explores the critical role of multi-stakeholder resilience building.

Disaster management and governments

Government, as a stakeholder and policymaker, has a pivotal role to play in the event of a disaster. The disaster management plans that are enforced by the government take a narrow approach as they are based on a dominant single perspective. This may have drawbacks for destination resilience building. One of the significant drawbacks of the prevailing disaster management plans is that they are dominated by 'restoration' approaches instead of assessing the existing system's relevancy and fitness through the eyes of multiple stakeholders. The haste to restore distracts from the opportunity disasters can offer to eliminate obsolete practices and integrate novel approaches that could be more relevant to stakeholders.

The effectiveness of disaster management plans and conceptualisations of disasters needs to be constantly monitored, reflecting the continually evolving nature of disasters and societies. This should include reassessing established disaster management plans, strategies and policies at each phase in light of a disaster (Negradas-Varona et al., 2017). By doing so, the best practices and errors in response to disasters can be identified, enabling a process of continual improvement. Such a reactive process is insufficient in itself. Rather, it must sit alongside a proactive process that continually reassesses plans and policies considering disasters and best practices elsewhere in the world, while recognising the need to situate potential lessons from these within the local context.

Disaster and disaster management knowledge is crucial to mitigate and respond to a disaster. Frequently, disaster knowledge is associated with creating awareness of potential disaster risks within a community (Odiase et al., 2020). In comparison, raising awareness of disaster knowledge amongst government officials, beyond disaster management experts, is seldom undertaken. It is vital to prepare emergency and disaster management officials through training and development (Balut et al., 2022), especially remembering the regular turnover of such people while looking to prepare others. Raising awareness of possible disasters and steps that need to be taken prior to and post-disaster amongst officials at all levels, no matter how directly or indirectly they are linked to a particular industry or governmental sector, is vital to the success of a disaster management plan. While training for disaster management is a vital tool in disaster preparedness, professionals within and outside the government regularly complain about inadequate training, teaching programmes and knowledge incorporation within their workplace (Naser & Saleem, 2018).

Disasters are dynamic, cross-disciplinary and multi-sectoral in their implications. They require officials, both disaster management experts and others, to collaborate across multiple sectors. The COVID-19 pandemic is a prime example in recent years of the complexity of disaster management. The epidemiology that once disrupted the world is now a matter that can be handled at the country level due to the joint efforts of disaster management and health experts, in tandem with those from virtually all other sectors of society. It is this cooperation that helps to control the intensity of a disaster (Williford, 2017). Regrettably, most governments do not encourage this approach, at least not in a proactive manner, due to many challenges. Engaging experts from across multiple disciplines and sectors and adopting new technologies is time-consuming and costly (Aguirre & El-Tawil, 2020). It also has the potential to trigger conflicts between disciplines and sectors (Bendito & Barrios, 2016).

The integration of power structures across local, provincial and national levels is key to the success of disaster management. Officials at each level should have a clear understanding of their roles and responsibilities. If this is not the case, disaster management plans and strategies have the potential to fail (Schneider, 2008). Communication at each level of the government and between disaster management authorities should be transparent to enable effective coordination (Yeh, 2021). Similarly, it is important to establish good coordination across inter-government agencies and departments. Countries following a federal government system often experience inter-government conflicts due to a lack of coordination. A lack of consideration of inter-governmental practices when designing disaster management policies is the leading cause of this challenge (Birkland & Waterman, 2008). This should be addressed in the mitigation and preparedness phases of disaster management to minimise the impact of a disaster.

The existence of different political perspectives is another challenge that should be managed to ensure departments within governments, links between them and their links to external sectors of society work effectively in times of crisis (Matsuura & Razak, 2019). Thus, effective disaster risk reduction requires minimal political interference on disaster management systems at times of crisis. Officials with different political perspectives tend to influence disaster management practices when they are implemented (Williams, 2011). This is mostly observed during the response and recovery phases of disaster management. Political biases are often apparent when governments and disaster management officials are involved with humanitarian support during a disaster. Occasionally, politicians tend to influence or interfere with those operationalising disaster management operations, interrupting ongoing rescue and relief efforts to direct aid for political gain (Dwivedi et al., 2018). This can assist politicians in acquiring new followers and maintaining a loyal voter base that could be beneficial in future elections (Bechtel & Hainmueller, 2011). A few politicians disrupt disaster management plans through corrupt practices for their own financial benefit. This is apparent when funds and goods received as aid are mismanaged. Politicians use state power to cover up such acts of corruption (Ha, 2023). Politicians in opposition parties view disasters as an opportunity to make political gains. They tend to reveal the mismanagement of disaster plans and strategies, and corruption, and often question the appropriateness and timeliness of government actions. These actions aim to change society's attitudes towards the ruling government (You et al., 2019). Whether such actions do anything to aid disaster management at the time of a crisis is very questionable. In all these cases there is the recognition that politicians are appealing to those who can vote and/or help fund their political parties and ultimately either keep them in power or help them gain power. Those who cannot do this (e.g., those young to vote, those from minority groups and those from lower socio-economic groups) may be of less interest.

Challenges faced by multi-stakeholders

Diverse challenges and grievances are present across multi-stakeholders at multi-levels regarding responding to disasters. The impact of disasters is felt by everyone, from the smallest tourist vendors to the largest multinational corporations (Samarathunga, 2020). While there are similarities, the way each stakeholder experiences a disaster is unique to them. Studies by UNDRR (2024b), Zorn (2018) and Khan et al. (2023) indicate that stakeholders in the least developed and developing countries are likely to experience a higher impact from disasters than those in developed countries. This is mainly due to their relatively vulnerable state, poor disaster management plans and strategies, and poor resilience planning and implementation. The capacity of each stakeholder at each level influences how disasters affect them. Capacity is referred to as the combination of stakeholders' strengths, resources and attributes available to manage and reduce disaster risks and strengthen resilience (UNDRR, 2024c). The ability to reestablish infrastructure and arrange finances and human capital to resume operations differs across stakeholders. Financial constraints are crucial as most of the businesses within the tourism industry are either small or micro. Approximately 80% of the businesses within tourism around the world are categorised as small- and medium-sized enterprises (SMEs) (World Travel & Tourism Council, 2022). These businesses operate with limited financial resources. The effects of disasters deteriorate their financial conditions even further. Policies imposed by governments tend to disregard the capacities and desires of multi-stakeholders at multiple levels as they typically take a top-down approach to disaster management. This leads multi-stakeholders to experience various challenges.

When a disaster strikes, destinations without strong governmental support structures provide the least direction to multi-stakeholders at the bottom and top levels. This can be more problematic for stakeholders who are not registered with the tourism authorities. Unregistered stakeholders have the potential to miss important communications issued by disaster management centres and tourism authorities. Crucial information, such as disaster warnings or information about rescue efforts, could be easily missed. Inaccurate information about stakeholders working in the grey economy has the potential to create a mismatch between the estimated and real information. This creates a conflict between disaster management implementations and reality. Tourism stakeholders at the bottom level, with limited awareness of policies and procedures, may have difficulties accessing information during and after a disaster. This has the potential to lead small- and medium-scale stakeholders to encounter system shock. A system shock is referred to as a sudden change that stakeholders are expected to cooperate with and adapt to in the immediate aftermath of a disaster (Scott et al., 2009). Any changes to disaster management strategies that occur during disasters create additional stress for SMEs as they are expected to cooperate despite their limited capabilities.

Businesses with vulnerable capacities, especially poor financial stability, tend to exit the tourism industry as a result of the impact of crises on them (Karunarathne et al., 2021). This can interrupt the supply chain of directly and indirectly linked businesses (Lu et al., 2020). Temporary or permanent business closure disrupts the flow of the tourism industry. Businesses adapting temporary closures find it difficult to resume operations, as when they attempt to re-open, they find their former employees have found alternative employment either within or outside the tourism industry (Quader et al., 2024). However, where closure, temporary or permanent, is a large-scale response to disasters employees face long-term unemployment (Blustein & Guarino, 2020). Families with sole income earners tend to suffer more. Loss of income leads people to frustration.

The effects of disasters can aggravate prevailing social challenges. These challenges exist at multiple levels for multi-stakeholders. In the aftermath of a disaster, increased unemployment is usually noticeable, the extent of it being governed by the scale of the disaster. This increases income inequalities in a destination. The loss of income due to the impact of disasters amongst wealthy groups tends to be high. Yet the relative impact on low-income earners and their families is far greater (UNDRR, 2023a).

Those in a vulnerable economic position before a disaster can be thrown into poverty by it. This can result in loss of shelter, something that is not just an individual's crisis but also a nations' crisis. People in poverty experience powerlessness and voicelessness (Narayan, 2000). They risk being mistreated by others due to a lack of strong policies protecting them. Multi-stakeholders across formal and informal institutes tend to communicate and interact with them less. Although people in poverty have a higher requirement of food, medicine, aid, better education and employment, they tend to receive the least (UNDRR, 2023a). These issues are experienced differently by various ethnicities.

Officials representing the dominant ethnic group in a destination have more potential to share their opinions and access information at each phase of disaster management. The use of a single language by governments inhibits the voices of other ethnic groups from being heard (Kassayie, 2006). This narrows the potential to incorporate wider solutions into disaster management plans. Also, it can increase the gap between the dominant ethnic group and the others (Bethel et al., 2013). Such discrimination and racism are often, rightly, identified as a spillover effect of colonialism (UN Human Rights Office, 2024) though colonialism alone is not responsible for it. Racism is not limited to private and public organisations (Griffith et al., 2007). Rather, it prevails at multi-levels and in every form of relationship, including professional and personal ones.

Racial and ethnic discrimination is linked to social stigma and cultural constraints. Linked to, but also separate from this is gendered discrimination. Such discrimination against women and sexual minorities is often seen in all parts of the world (Jayachandran, 2015). This discrimination places women in vulnerable positions, something only worsened during disasters due to their enforced lack of involvement and prominence in disaster management. These conditions are worse for women in many developing countries and rural areas, and especially for those with a tourism entrepreneurial mindset. Some cultures view women taking part in a business as socially inappropriate and owning and running one as even worse. However, in such destinations, women are often silently involved in these practices, an economic necessity that society turns a blind eye to. In such societies, SME accommodation operations often place another burden on women on top of the traditional household work demanded of them. They are expected to cook, clean and do laundry for the tourists without any direct financial gain or recognition for their efforts. They may also run the finances of the business and any IT functions and manage the operation. In such destinations, women are very rarely involved in emergency preparedness and disaster management even though they run the internal operations of the tourist business and are actually the 'experts'. Instead, they often miss out on early warning information and details about rescue efforts because of societal expectations associated with gender. Furthermore, in a post-disaster scenario, women tend to receive the least access to aid (UNDP, 2023) or, at best, receive it under the label of 'victims' rather than active social agents capable of driving change themselves. This indicates that women stakeholders experience diverse and unique gendered challenges at multiple levels. Also, this implies the need to look at formulating policies that strengthen the position of women within tourism and disaster management.

Most of the challenges experienced by multi-stakeholders across multiple levels are not newly discovered, yet they often remain unsolved in many locations. This relates to multiple reasons, including:

I The inability of policymakers to recognise the real challenges that are faced by multi-stakeholders across multiple levels.
II Government and associated officials have a bird's eye view of the incidents taking place at different levels.
III The focal point of governments keeps shifting to other or new incidents.
IV Limited capacities of governments.
V Perspective of personnel that are involved in policy formation and development.
VI Purposefully disregarding and prioritising other things over these challenges depending on the context and capacity of the government.

These reasons and the challenges faced by multi-stakeholders outlined above indicate there is a serious gap within the prevailing disaster management and tourism policies. The continuity of these challenges across multiple levels indicates that the policies often imposed by governments tend to lack a multi-stakeholder perspective. Hence, destinations often struggle to recover their tourism industry post-disaster, and even when success appears to occur, it can hide a more nuanced reality that sees many left suffering and/or unable to take advantage of emergent opportunities. The duration of the recovery period indicates the strengths and weaknesses of the strategies and policies that are in place to respond to a disaster. It is important to recognise that such a duration is not homogenous within a destination.

Is resilience building possible alone?

It is impossible for governments to eradicate all challenges, yet it is critical that they form policies that address the majority and enhance resilience. Yet it is also important that they do so in a way that does not favour one group over another. To do this it is essential to look at disaster management and tourism beyond a singular perspective. Giving a multi-stakeholder perspective prominence can improve engagement and coordination to form policies that are relevant to all. This can ignite a sense of inclusion, ownership and shared responsibility. Policies taking a dynamic, multi-stakeholder and inclusive approach provide an opportunity to bring new ideas and solutions to crises from multiple angles. This also reduces the degree of government imposition and improves the cohesiveness amongst all stakeholders, including the government. Thus, it is important to build resilience by incorporating multi-stakeholder perspectives. This requires synchronisation and empowerment across multi-stakeholders and multiple levels.

Cohesiveness across stakeholders presents multiple paths to help build resilience in the face of disasters. It assists in identifying the real strengths and weaknesses of each stakeholder and how they can collectively help one another. This provides the opportunity to reassess and identify the fitness of existing disaster management and tourism policies. While cohesive efforts have the potential to minimise vulnerabilities, they also provide an opportunity to capitalise on the strengths of different stakeholders to support others across various levels. This demands a different approach to neo-liberalism that venerates the success of the individual at the expense of the other.

Cross-learning from different regions within and outside a destination has the potential to improve existing strategies within disaster management and tourism. Also, it improves communication amongst countries, regions and the world. Better communication has the potential to alert other countries of a possible disaster and how to deal with it collectively. This information has the potential to reduce the impact of a disaster across destinations. This can help to improve international relations. This way, stakeholders may benefit not just from relevant disaster management policies and strategies but also from regional tourist business opportunities. Business collaboration, foreign investment and tourist product imports and exports are a few opportunities related to this. When disaster knowledge is shared across borders, regional tourists/travellers have the potential to travel more after a disaster as they are more aware of destination conditions. Cohesive effort across all stakeholders does not just strengthen the tourism industry. Instead, it strengthens other associated social, cultural, economic and environmental elements within and across countries.

Conclusion

Many tourist destinations across the world have experienced and will continue to experience multiple natural and man-made disasters. As highlighted in the above sections, the prevailing tourism disaster management policies that are led by governments often fail to address many of the challenges faced by multi-stakeholders at multiple levels. This, along with the increasing number of disasters and their scale, has an increasingly detrimental effect on the tourism industry and everyone associated with it. Therefore, destination resilience building has become critical.

This chapter has emphasised the necessity of looking at tourism disaster management policies beyond a single perspective. Instead, a way of thinking and acting that is relevant to all stakeholders at all levels is needed. The strategies and policies at each phase of disaster management should take an all-inclusive approach to reduce the implications of disasters and strategies with a dominant approach. This could be facilitated through cohesive efforts, coordination, collaboration and communication across not just a few stakeholders but all stakeholders at each level. The empowerment of the other through this process is ultimately beneficial for all in a way that looks beyond the individualism of neo-liberalism.

As the first step towards such an approach, policymakers need to recognise the primary gaps in their existing policies and strategies in each disaster management phase. This is possible by identifying the key challenges faced by all stakeholders. In particular, stakeholders at the bottom level should be focused on as their vulnerable capacities provide limited scope to prepare better for a disaster and rise post-disaster.

It is vital to develop relationships with stakeholders at multiple levels who are directly and indirectly associated with the tourism industry. These relationships need to see beyond ingrained discriminatory cultures. This provides the opportunity to recognise the importance of disaster risk through the lens of multi-stakeholders at multiple levels. Disaster management plans should focus on preparing all stakeholders for possible impacts. Furthermore, strategic planning and implementations should enhance resilience building of multi-stakeholders, recognising that, in turn, this aids destination resilience building.

As we live in a world dominated by neo-liberalism, establishing a multi-stakeholder approach across multiple levels is challenging. Most people seek advantage at the expense of others, constantly seeking an edge, an advantage, something that can be exploited. While this is recognised as a barrier to a multi-stakeholder approach, it is vital that tourism disaster management policies take a holistic approach for effective destination resilience building.

References

Aguirre, B., & El-Tawil, S. (2020). The Emergence of Transdisciplinary Research and Disaster Science. *American Behavioral Scientist, 64*(8), 1162–1178. https://doi.org/10.1177/0002764220938114

Asian Disaster Reduction Center. (2022). *Total Disaster Risk Management – Good Practices*. Asian Disaster Reduction Center. Retrieved June 20, 2024, from https://www.adrc.asia/publications/annual/09/09eng/pdf/5-3.pdf

Azad, M., Uddin, M., Zaman, S., & Ashraf, M. (2019). Community-Based Disaster Management and Its Salient Features: A Policy Approach to People-centred Risk Reduction in Bangladesh. *Asia-Pacific Journal of Rural Development, 29*(2), 135–160. https://doi.org/10.1177/1018529119898036

Balut, M., Der-Martirosian, C., & Dobalian, A. (2022). Disaster Preparedness Training Needs of Healthcare Workers at the US Department of Veterans Affairs. *Southern Medical Journal, 115*(2), 158–163. https://doi.org/10.14423%2FSMJ.0000000000001358

Bechtel, M., & Hainmueller, J. (2011). How Lasting Is Voter Gratitude? An Analysis of the Short- and Long-Term Electoral Returns to Beneficial Policy. *American Journal of Political Science, 55*(4), 852–868. https://doi.org/10.1111/j.1540-5907.2011.00533.x

Bendito, A., & Barrios, E. (2016). Convergent Agency: Encouraging Transdisciplinary Approaches for Effective Climate Change Adaptation and Disaster Risk Reduction. *International Journal of Disaster Risk Science, 7*, 430–435. https://doi.org/10.1007/s13753-016-0102-9

Bethel, J., Burke, S., & Britt, A. (2013). Disparity in Disaster Preparedness Between Racial/Ethnic Groups. *Disaster Health, 1*(2), 110–116. https://doi.org/10.4161%2Fdish.27085

Birkland, T., & Waterman, S. (2008). Is Federalism the Reason for Policy Failure in Hurricane Katrina? *Publius, 38*, 692–714. https://www.jstor.org/stable/20184998

Blustein, D. L., & Guarino, P. A. (2020). Work and Unemployment in the Time of COVID-19: The Existential Experience of Loss and Fear. *Journal of Humanistic Psychology, 60*(5), 702–709. https://doi.org/10.1177/0022167820934229

Center For Disaster Philanthropy. (2023). *Disaster Phases.* Retrieved June 20, 2024, from Center For Disaster Philanthropy: https://disasterphilanthropy.org/resources/disaster-phases/

CRED. (2024). 2023 Disasters in Numbers. Centre for Research on the Epidemiology of Disasters (CRED). Retrieved June 29, 2024, from https://reliefweb.int/report/world/2023-disasters-numbers

Das, P. S. (2023). Good Governance Strategies for Disaster Management and Risk Reduction. In A. Singh (Ed.), *International Handbook of Disaster Research* (pp. 2097–2112). Singapore: Springer. https://doi.org/10.1007/978-981-19-8388-7_145

Dwivedi, Y., Shareef, M., Mukerji, B., Ranaa, N., & Kapoor, K. (2018). Involvement in Emergency Supply Chain for Disaster Management: A Cognitive Dissonance Perspective. *International Journal of Production Research, 56*(21), 6758–6773. https://doi.org/10.1080/00207543.2017.1378958

Griffith, D., Childs, E., Eng, E., & Jeffries, V. (2007). Racism in Organizations: The Case of a County Public Health Department. *Journal of Community Psychology, 35*(3), 287–302. https://doi.org/10.1002%2Fjcop.20149

Ha, K.-M. (2023). Decreasing Corruption in the Feld of Disaster Management. *Crime, Law and Social Change, 80*, 195–214. https://doi.org/10.1007/s10611-023-10077-y

Jayachandran, S. (2015). The Roots of Gender Inequality in Developing Countries. *Annual Review of Economics, 7*(1), 63–88. https://doi.org/10.1146/annurev-economics-080614-115404

Karunarathne, A., Ranasinghe, J., & Sammani, U. (2021). Impact of the COVID-19 Pandemic on Tourism Operations and Resilience: Stakeholders' Perspective in Sri Lanka. *Worldwide Hospitality and Tourism Themes, 13*(3), 369–382. https://doi.org/10.1108/WHATT-01-2021-0009

Kassayie, B. (2006). Urban Multilingualism and the Development of a Local Communication Support Strategy in Tower Hamlets: Fieldwork report. *International Journal of Migration, Health, and Social Care, 2*(1), 27–47. https://doi.org/10.1108/17479894200600004

Khan, M., Anwar, S., Sarkodie, S., Yaseen, M., & Nadeem, A. (2023). Do Natural Disasters Affect Economic Growth? The Role of Human Capital, Foreign Direct Investment, and Infrastructure Dynamics. *Heliyon, 9*(1), 1–19. https://doi.org/10.1016/j.heliyon.2023.e12911

Lu, Y., Wu, J., Peng, J., & Lu, L. (2020). The Perceived Impact of the Covid-19 Epidemic: Evidence from a Sample of 4807 SMEs in Sichuan Province, China. *Environmental Hazards, 19*(4), 323–340. https://doi.org/10.1080/17477891.2020.1763902

Matsuura, S., & Razak, K. (2019). Exploring Transdisciplinary Approaches to Facilitate Disaster Risk Reduction. *Disaster Prevention and Management, 28*(6), 817–830. https://doi.org/10.1108/DPM-09-2019-0289

Mushkatel, A., & Weschler, L. (1985). Emergency Management and the Intergovernmental System. *Public Administration Review, 45*(01), 49–56. https://doi.org/10.2307/3134997

Narayan, D. (2000). Poverty is Powerlessness and Voicelessness. *Finance and Development, 37*(4), 18–21. https://doi.org/10.5089/9781451951936.022

Naser, W., & Saleem, H. (2018). Emergency and Disaster Management Training; Knowledge and Attitude of Yemeni Health Professionals – A Cross-sectional Study. *BMC Emergency Medicine, 18*, 1–12. https://doi.org/10.1186/s12873-018-0174-5

Negradas-Varona, R. N., Aya, M.D., Bolla, H., Bolinget M. S., & Illab, H. S. (2017). Knowledge, Attitude and Practices on Disaster Risk Reduction and Management of the Barangay Officials of Baler, Aurora, Philippines. *International Journal of Advanced Research (IJAR)*, 5(7), 1395–1402. http://dx.doi.org/10.21474/IJAR01/4851

Noy, I. (2015, March 16). How Can We Measure the Impact of Natural Disasters? Retrieved July 10, 2024, from World Economic Forum: https://www.weforum.org/agenda/2015/03/how-can-we-measure-the-impact-of-natural-disasters/

Odiase, O., Wilkinson, S., & Neef, A. (2020). Risk of a Disaster: Risk Knowledge, Interpretation and Resilience. *Jàmbá – Journal of Disaster Risk Studies, 12*(1), 1–9. https://doi.org/10.4102/jamba.v12i1.845

Quader, M., Hossain, M., & Hassan, H. (2024). Stakeholders' Views about Consequences of COVID-19 Epidemic on the Tourism Industry of Bangladesh: Reconciliation Policy Framework. *Cogent Social Sciences, 10*(1), 1–18. https://doi.org/10.1080/23311886.2024.2318869

Rossello, J., Becken, S., & Santana-Gallego, M. (2020). The Effects of Natural Disasters on International Tourism: A Global Analysis. *Tourism Management, 79*, 1–11. https://doi.org/10.1016%2Fj.tourman.2020.104080

Samarathunga, W. (2020). Post-COVID19 Challenges and Way Forward for Sri Lanka Tourism. *Social Science Research Network (SSRN)*, 1–12. https://doi.org/10.2139/ssrn.3581509

Schneider, S. (2008). Who's to Blame? (Mis) Perceptions of the Intergovernmental Response to Disasters. *Publius, 38*, 715–738. https://www.jstor.org/stable/20184999

Scott, D., de Freitas, C., & Matzarakis, A. (2009). Adaptation in the Tourism and Recreation Sector. In K. Ebi, I. Burton, & G. McGregor (Eds.), *Biometeorology for Adaptation to Climate Variability and Change. Biometeorology* (Vol. 1, pp. 171–194). Dordrecht: Springer. https://doi.org/10.1007/978-1-4020-8921-3_8

Smith, W., & Carmichael, B. (2006). Canadian Seasonality and Domestic Travel Patterns: Regularities and Dislocations as a Result of the Events of 9/11. *Journal of Travel & Tourism Marketing, 19*(2), 61–76. https://doi.org/10.1300/J073v19n02_06

UN Human Rights Office. (2024). *Office of the United Nations High Commissioner for Human Rights*. Retrieved May 09, 2024, from Racism, discrimination are legacies of colonialism: https://www.ohchr.org/en/get-involved/stories/racism-discrimination-are-legacies-colonialism

UNDP. (2023). *Disaster Recovery Framework Guide for the Tourism Sector*. New York: United Nations Development Programme. Retrieved June 20, 2024, from https://www.preventionweb.net/publication/tourism-sector-disaster-recovery-framework-guide

UNDRR. (2022). *Global Assessment Report on Disaster Risk Reduction (GAR) 2022: Our World at Risk: Transforming Governance for a Resilient Future*. Geneva: United Nations Office for Disaster Risk Reduction. Retrieved June 20, 2024, from https://www.undrr.org/gar/gar2022-our-world-risk-gar

UNDRR. (2023a). *Poverty and Inequality*. Retrieved March 07, 2024, from United Nations Office for Disaster Risk Reduction: https://www.preventionweb.net/understanding-disaster-risk/risk-drivers/poverty-inequality#:~:text=Recovery%20after%20disaster%20can%20extremely,them%20to%20even%20greater%20risk

UNDRR. (2023b). *Sendai Framework Terminology on Disaster Risk Reduction: Mitigation*. Retrieved June 20, 2024, from United Nations Office for Disaster Risk Reduction: https://www.undrr.org/terminology/mitigation

UNDRR. (2023c). *Sendai Framework Terminology on Disaster Risk Reduction: Preparedness*. Retrieved June 20, 2024, from United Nations Office for Disaster Risk Reduction: https://www.undrr.org/terminology/preparedness#:~:text=The%20 knowledge%20and%20capacities%20developed,likely%2C%20imminent%20or%20 current%20disasters

UNDRR. (2023d). *Sendai Framework Terminology on Disaster Risk Reduction: Recovery*. Retrieved June 20, 2024, from United Nations Office for Disaster Risk Reduction: https://www.undrr.org/terminology/recovery

UNDRR. (2024a). *Disaster*. Retrieved June 21, 2024, from United Nations Office for Disaster Risk Reduction: https://www.undrr.org/terminology/disaster

UNDRR. (2024b). *Disaster Risk Reduction in Least Developed Countries*. Retrieved July 10, 2024, from United Nations Office for Disaster Risk Reduction: https://www. undrr.org/implementing-sendai-framework/sendai-framework-action/disaster-risk-reduction-least-developed-countries#:~:text=Impact%20of%20disasters%20on%20 LDCs&text=Disasters%20impact%20the%20economies%20of,Displacement%20 Monitoring%20Centre%2C%

UNDRR. (2024c). *Sendai Framework Terminology on Disaster Risk Reduction: Capacity*. Retrieved July 06, 2024, from United Nations Office for Disaster Risk Reduction: https://www.undrr.org/terminology/capacity

UNESCO. (2023, June 09). *Disaster Risk Reduction: Post-Disaster Response*. Retrieved June 20, 2024, from United Nations Educational, Scientific and Cultural Organization: https://www.unesco.org/en/disaster-risk-reduction/post-disaster

UNISDR. (2012). *Disaster Risk and Resilience*. United Nations. Retrieved June 30, 2024, from https://www.un.org/en/development/desa/policy/untaskteam_undf/ thinkpieces/3_disaster_risk_resilience.pdf

Williams, G. (2011). *The Political Economy of Disaster Risk Reduction: Study on Disaster Risk Reduction, Decentralization and Political Economy*. Brighton, United Kingdom: UNDP Bureau for Crisis Prevention and Recovery (BCPR). Retrieved June 15, 2024, from https://www.preventionweb.net/english/hyogo/gar/2011/en/ bgdocs/Williams_2011.pdf

Williford, D. (2017). Seismic Politics: Risk and Reconstruction after the 1960 Earthquake in Agadir, Morocco. *Technology and Culture, 58*, 982–1016. https://doi. org/10.1353/tech.2017.0111

World Travel & Tourism Council. (2022). *Travel & Tourism Economic Impact 2022*. World Travel & Tourism Council. Retrieved June 30, 2024, from https://wttc.org/ research/economic-impact

Yeh, S.-S. (2021). Tourism Recovery Strategy against COVID-19 Pandemic. *Tourism Recreation Research, 46*(2), 188–194. https://doi.org/10.1080/02508281.2020.1805933

You, Y., Huang, Y., & Zhuang, Y. (2019). Natural Disaster and Political Trust: A Natural Experiment Study of the Impact of the Wenchuan Earthquake. *Chinese Journal of Sociology, 6*(1), 140–165. https://doi.org/10.1177/2057150x19891880

Zorn, M. (2018). Natural Disasters and Less Developed Countries. In S. Pelc, & M. Koderman (Eds.), *Nature, Tourism and Ethnicity as Drivers of (De) Marginalization* (Vol. 3, pp. 59–78). Springer International Publishing. https://www. springerprofessional.de/en/natural-disasters-and-less-developed-countries/13342842

11 Defining Resilience Capabilities in Travel Systems

Wong Ling Chai, Siao Fui Wong, Bee Lian Song,
Tee Poh Kiong, Ying Ying Tiong and
Lim Ming Fook

11.1 Introduction

The travel industry encompassed hotels, airlines, tour operators, travel agencies, and destination marketing organisations (Nour El-Din et al., 2023). As an industry with a diverse range of companies that focus on meeting the unique needs of travellers (Camilleri & Camilleri, 2018), the travel industry contributes significantly as job suppliers in the global economy that increases GDP in many developing countries (Sinclair, 1998). However, the growth of tourism is closely tied to environmental pollution (Xiong et al., 2023). Different sectors within the tourism and travel industry have varying effects on economic growth and pollution levels.

Other than environmental impacts, businesses operating in the global market also face a range of challenges including changing consumer preferences, technological progress, and globalisation effects (Prasanna et al., 2019). Additional elements contributing to tourism vulnerability are geopolitical tensions, economic crises, technical breakthroughs, and environmental concerns (Harrington, 2021). These vulnerabilities can impact workers, employers, and visitors, resulting in exploitation, inadequate wages, extended working hours, and discrimination against individuals with disabilities (Isichei, 2022).

Vulnerability is most pronounced among low-skills and less advantaged individuals. It differs across geographical areas, with rural and remote areas experiencing distinct vulnerabilities owing to restricted resilience options (Marcouiller, 2023). In a similar vein, the tourism sector is vulnerable to a range of disturbances such as crises, disasters, and other disruptive occurrences. These interruptions can have dual effects on tourism, both beneficial and detrimental (Kaur & Noonwal, 2023). To address the aforementioned vulnerabilities, it is important to have adequate policies in place to protect the vulnerable and enhance the resilience capabilities of the travel system.

Resilience in the tourism industry refers to the ability of tourism stakeholders to survive and recover from natural disasters or calamities (Prayag, 2023). The process resilience offers risk mitigation alternatives, adversity adaptation, and effective solutions for the travel system (Mitchell, 2013). These include engaging multiple stakeholders, perceiving tourism sites through complex

DOI: 10.4324/9781032720555-12

systems, communicating in a common language, and proactively managing risks (Pennington-Gray & Basurto-Cedeno, 2023). The relevance of resilience to travel systems is portrayed through approaches to respond, adapt, and survive in times of adversity (Hall et al., 2018).

Travel systems refer to the coordination and management of transportation and information within the tourism industry. It involves the effective movement of people, goods, and vehicles in order to reduce costs and improve delivery times (Dileep & Pagliara, 2023). Dar et al. (2022) highlighted the significance of the marketing information system to the travel system, stressing its role in the collection, analysis, and dissemination of pertinent information that is used to enhance the planning, implementation, and control of tourism marketing activities. The travel system comprises various components, including transportation, information management, marketing strategies, and itinerary management, aimed to enhance the overall travel experience for tourists. It provides both static and dynamic travel itineraries, enabling real-time monitoring of travel plans and changing of itineraries (de Moraes Ramos et al., 2020). Owing to the digital revolution, the tourism industry has been completely transformed, resulting in the development of new tourism consumption models.

Resilience is important to travel systems as guides for tourism stakeholders to maintain or quickly resume operations after disruptions (Bec et al., 2016). Resilience is frequently measured using metrics and mathematical models, and methods for improving it are constantly being investigated (Sharma et al., 2018). Scholars (Prayag, 2023; Prayag et al., 2024a; Prayag et al., 2024b) even discovered four levels of resilience in the travel system: individual, organisational, community, and destination. However, more cross-disciplinary collaboration is required to turn research findings into practical strategies (Mattsson & Jenelius, 2015). Although existing studies have provided insights into resilience capabilities building (Prayag, 2023; Prayag et al., 2024a; Prayag et al., 2024b), there remain limited guidelines or frameworks available to stakeholders or policymakers to react in such instances. To fill this gap in the literature, this chapter attempts to define resilience capabilities in travel systems and offer practical implications guiding stakeholders and policymakers.

11.2 The Concept of Resilience in Travel Systems

Osorio et al. (2023) define resilience as a person's or organisation's ability to mitigate risk or achieve stability again, including the capacity to take in, recover from, and adapt to various known and unknown risks. The term "resilience" has gained popularity in the hospitality and tourism fields to describe the ability of the tourism system – which includes the industry, destinations, communities, tourism operators, homestays as well as individuals (such as., tourists) – to deal with crises (Cheer & Lew, 2017; Hall et al., 2018; Prayag, 2018) and respond to change in a way that preserves stability while allowing for innovation and future development (Kaushal & Srivastava, 2021). Previous literature on resilience focused primarily on fields such as psychology, marketing, and change and disaster management; however, there is a growing body of research on destination resilience, which focuses on a person's behaviour during times of adversity (Hall et al., 2018; Prayag et al., 2020). As a result, the literature on psychological resilience served as an underpinning for the conceptualisation of resilience of the tourism system in this study.

Psychological resilience is described as a personal quality or behavioural capacity to rely on social supports and develop resources that enable a person to confront unforeseen challenges and adapt to adverse circumstances (Hall et al., 2018). Respectively, the resilience concept is framed as an attribute or a capability inclusive of the notion of preparedness, adaptation, redundancy, recovery, and flexibility emerge through different uses and applications of the concept across discipline and context (Cheer & Lew, 2017). According to Prayag et al. (2024a), previous experience with crises and change may enable organisations to learn and respond appropriately to new challenges and crises. As a result, the ability of firms, destinations, and individuals to demonstrate resilience has become recognised as a source of competitiveness (Hall et al., 2018). Understanding the notion of resilience is particularly important for scholars as well as tour operators since research is still ambiguous concerning the adaptability and the learning capability to translate the past experience into new and emerging crises (Bhaskara & Filimonau, 2021; Ghaderi et al., 2022). According to Hassler and Kohler (2014, p. 119), the concept of resilience can be ambiguous, contradictory, and open to interpretation.

Adaptability, robustness, redundancy, and flexibility are key features of flexibility in travel systems, namely flexible systems such as connectivity and modularity (Masik, 2022). Adaptability, as emphasised by Feng et al. (2023), shows the ability of the system to respond and recover from uncertainties, reflecting changes in passenger behaviour. Robustness and redundancy, as reported by Du et al. (2022), are prerequisites for violence that ensure a smooth operation according to the system's definition and maintain service delivery through two channels and other assets. Flexibility, a key asset in resilient systems (Masik, 2022) allows for a flexible response to disturbances, while capacity utilisation and speed (Du et al., 2022) have post-perturbation properties that enable early detection of performance problems and rapid recovery of both energy and channels. All these considerations contribute to increasing the resilience of travel systems to various challenges.

In recent years, the overall tourism eco-system has been severely impacted by the COVID-19 pandemic. According to the United Nations World Tourism Organisation (UNWTO, 2020), the pandemic has negatively impacted not merely the economic growth of tourism industries, but also social impacts by means of loss of employment and an increase in the risk perception of potential tourists concerning destination safety and security. It is important to note that tourism recovery may be delayed if tourism management organisations are unable to adequately prepare for and adapt to changes (Hall et al., 2018). As a result, resilience, much like sustainability, is frequently portrayed as an important approach for tourism and hospitality businesses, organisations, and communities to respond, adapt, and survive change (Hall et al., 2018; Pocinho et al., 2022). Indeed, scholars in the field of change and crisis management advocate for destinations to become more resilient in order for businesses or industries to recover and return to "normality" (Scott & Laws, 2006). The post-COVID implications for the tourism and hospitality industries have drawn more attention to the concept of resilience than ever before (Bhaskara & Filimonau, 2021). Therefore, this study examines the application of the concept of resilience in hospitality and tourism, specifically investigates the factors influencing resilience in travel systems and the development of resilience capabilities through travel systems-emphasised factors.

11.3 Factors Influencing Resilience in Travel Systems

Prayag (2023) categorised tourism system resilience into four levels: individual level (psychological, employee, entrepreneurial, tourist, and resident resilience); organisational level (team, supply chain, and business cluster resilience); community level (community networks, NGOs, local government, and temporary visitor population resilience); and destination level (intraregional, national, and transnational resilience). Individual and organisational levels are lower-level tourism systems that can be explained with ecological and engineering perspectives while community and destination levels are higher-level systems better explained with socio-ecological systems (Prayag, 2023). Each level of resilience has different system characteristics and is influenced by various factors, but characteristics and factors of lower-level resilience can be applied to higher-level resilience (Prayag, 2023). Thus, factors influencing each level are discussed from low to high levels of the tourism system.

Individual level resilience denotes the capabilities of individuals, employees, entrepreneurs, tourists, and residents to cope with adversity. Psychological (individual) resilience is an individual's ability to leverage his/her own strengths and weaknesses to adapt and cope with adversity (Hall et al., 2018). Psychological resilience exists in various contexts (i.e., at workplace, at home); therefore, its influencing factors vary across different circumstances. Panzeri et al. (2021) applied the ecological resilience model and found personality traits (neuroticism, conscientiousness) promote resilience while other factors (i.e., loneliness) hinder resilience. Employee resilience is an employee's ability to use organisational resources to adapt, cope, and thrive in times of adversity (Nguyen et al., 2016). Providing organisational resources and support can help employees leverage their coping capacity (Hall et al., 2018). Employee resilience is influenced by psychological resilience because resilient individuals often hold positive attitudes towards work and life (Prayag et al., 2020). Entrepreneur resilience is an entrepreneur's ability to adapt and recover from challenges in their entrepreneurial venture (Prayag, 2023). Prayag (2023) differentiated entrepreneur resilience from psychological resilience by highlighting the importance of entrepreneurial thinking and eco-system support in an entrepreneur's decision-making process. Tourist resilience is a tourist's ability to use psychological and social resources to adapt and cope with adversity (Gottschalk et al., 2022). It differs from psychological resilience in its contribution to destination resilience (Hall et al., 2018). Gottschalk et al. (2022) proposed that tourist preparedness, adaptiveness, social support, and risk perceptions influence their resilience. Resident resilience is a poorly theorised concept understood as a resident's ability to adapt and cope with tourism (re) development (Prayag, 2023). Its influencing factors include social, economic, infrastructure, community capital, and institution but are strongly affected by tourism demand that enhances community capacity to respond to tourism (re) development or changes (Yang et al., 2021).

Organisational level resilience is an organisation's adaptability and tolerance to disruption and ability to unlock opportunities from the changing environment (Hall et al., 2018). Organisational resilience is determined by various factors. Hall et al. (2018) and Prayag et al. (2020) used the socio-ecological system to explain the role of employee resilience in organisational resilience post-disaster. Organisational level resilience consists of three understudied concepts, namely team, supply chain, and business cluster resilience (Prayag, 2023). Team resilience is a team's adaptability to adversity resulting from processes fostered through team composition and contextual factors, such as transformational leadership, team member relationships, and team culture (Hartwig et al., 2020). Team resilience is understudied in tourism and hospitality, but its contribution to organisational success and resilience is well-recognised (Prayag, 2023). Supply chain resilience is the ability of a supply chain to reduce, resist, recover, and react to disruptive events and restore its state of operations (Aigbedo, 2021). From the engineering perspective, improving supply chain agility and forming collaborative supply chain relationships can enhance supply chain resilience (Tukamuhabwa et al., 2015). Business cluster resilience is the adaptability of same-sector organisations to adversity (Prayag, 2023). Despite not being theorised in tourism and hospitality literature, Prayag (2023) highlights the importance of collaboration between the same-sector and bordering organisations and suggests number of organisations, distance between organisations, and socio-cultural relationships within organisation networks may affect business cluster resilience.

Community-level resilience is community preparedness and responsiveness to adversity and the capability to recover from it (Yang et al., 2021). Community resilience is influenced by tourism demand as it enhances the local economy (Yang et al., 2021). To improve community resilience, some small islands in developing states are encouraging tourist-involvement activities like tree-planting or turtle conservation (Kampel, 2022), such as the Turtle Islands Park in Sabah, Malaysia. Community resilience in tourism and hospitality includes community networks, NGOs, local government, and temporary visitor population resilience (Prayag, 2023). Collaboration among community networks (tourism and non-tourism agencies), NGOs, and local government can strengthen community resilience while network relationships and social learning can influence temporary visitor population resilience (Prayag, 2023).

Self-organisation and dependency are also important influencing factors of community resilience. For example, the growth of tourism in the Miao communities in Hunan Province, China, has not only diversified the local economy but enhanced community resilience through which villagers survived the epidemic by maintaining small-scale agricultural activities (Tian et al., 2023).

Destination level resilience occurs in three contexts namely intraregional, national, and transnational (Prayag, 2023). From the socio-ecological system perspective, strengthening collaboration practices and networks at the individual, organisational, and community levels can improve intraregional resilience (Prayag, 2023). For example, New Zealand responded to tourists' increasing consciousness of environmental impact by resetting tourism post-COVID-19 with sustainable and cleaner methods of consumption (Kampel, 2022). At the national level, resilience is influenced by the products and competitive advantages of destinations or regions (Prayag, 2023). This indicates that collaborative destination marketing, referring to collaborative practices between public and private organisations and tourists (Wong & Kler, 2022), would enhance destination resilience in tourism. At the transnational level, resilience is affected by funding and support (Prayag, 2023). For instance, Finland's government allocated 4 million euros to fund the tourism industry and region recovery in 2020; Canada provided 4 billion Canadian dollars for SMEs including the tourism sector to develop digital technologies; and Greece provided support for investments in digital technologies and services (OECD, 2022). Transnational resilience is also influenced by the consequences of war as it raises inflation and slows down the economy in affected countries (OECD, 2022).

11.4 Measuring Resilience Capacities in Travel Systems

Resilience in the tourism industry is an important factor to stay competitive in the market. Bui and Wickens (2021) observed three resilience levels as standards of actual resilience activities among tourism businesses. The first level ("macro-level") focuses on two key elements: tourism reshaping and tourism research agenda resetting; and future tourism industry redesigning. The second level ("meso-level") highlights organisational resilience in hotels; the tourism industry, research and education resetting; and local food systems resilience. The third level ("micro-level") emphasises travel fear and employee's psychological capital resilience. The first level is imperative as it depends on international arrival, whereas the second level focuses more on rebuilding the industry at the domestic level and puts emphasis on the travel system resilience. Micro-level, on the other hand, shows the impacts towards the stakeholders, suggesting both psychological capital resilience and employees' policymaking should be adapted at the macro- and meso-level of tourism rebuilding (Bui & Wickens, 2021).

Biggs' (2011) study on reef tourism revealed that the sector is vulnerable to both ecological effects and economic conditions. The findings were in line with earlier evidence in the 90s that healthy financial conditions are key to business survival and success (Bates, 1990). It is still relevant even after COVID-19, tourism players with high cash-driven resilience capabilities have earned higher profits and are less financially constrained (Wieczorek-Kosmala, 2022). Another fun fact observed by Wieczorek-Kosmala (2022) is that this phenomenon only differs among firms with different business sizes, yet shows no difference between firms across different countries. The findings indicate the importance of cash possession for tourism players and financial slack in building resilience capabilities.

Despite the context difference, the consistency introduced above is somehow supported by Prayag et al. (2024b), who explored the difference between planned and adaptive resilience of numerous travel sectors, including tourism, hospitality, and aviation. In their study, a consistent pattern was found in planned resilience, represented by change resilience over adaptive resilience, which describes leadership and culture, networks and relationships at all crisis stages. The study further analysed resilience types in stages: short-term and long-term responses. The short-term response stage highlighted discrepancies in priority among the travel sectors, hotels prioritise recovery and staff engagement, and planning strategies are the top priority of the restaurant and aviation sectors. The long-term recovery stage shows alignment between all sectors, emphasising business resilience-building strategies (change of routine, updating service, and developing new products).

Resilience, though having diverse definitions, *"is a capacity to resist being 'put down,' and the ability to recover and thrive from traumatic situations"* (Harms et al., 2018). Understanding resilience post-pandemic will help practitioners in rebuilding their capabilities effectively.

11.5 Relevant Case Studies

Resilient travel systems are critical to achieve business growth and sustainability. This section provides selected case studies of various companies across the globe that have practised resilient travel systems.

a **Singapore Airlines (Singapore)** – Singapore's flag carrier airline and ranked among the world's best airlines. Known for its quality service standards, the airline has enhanced its multi-hub strategy through more effective and sustainable partnerships. Through its established networks in Star Alliance, it deepened its ties with Southeast Asian airlines (i.e., Garuda Indonesia, Malaysia Airlines) and partnered with Tata Sons of India to achieve higher traffic to respective hubs (SIA, 2022; SIA, 2023). Singapore Airlines has taken fast decision approach to ensure its immediate survival while focusing on its long-term strategic goal to restore growth and profitability during and post-pandemic period. Furthermore, Singapore Airlines leveraged on more effective resilient travel systems through sophisticated digital systems providing contactless options for customers, service digital upgrades in their inflight entertainments, and health checks.

b **Huawei Enterprise (China)** – a company that offers innovative information and communications technology infrastructure products and solutions for various industries. Emphasising sustainable digital and intelligent transformation strategies, Huawei has contributed to the resilient travel system in the aviation industry through five end-to-end capabilities: full connectivity mainly for network, cloud-based service, data convergence and refinement, platform capability, and intelligence application. The fully connected network emphasises health and safety through enhanced planes' facilities cleaning procedures and contactless processing. The network's flexibility, such as pay-per-use in modern cloud technology, contributes to higher efficiency in operations. Huawei's Smart Airport Solution is a digital platform that leverages artificial intelligence (AI), the internet of things (IoT), big data and video cloud streamlines passenger and flight flows, airport operational efficiency and security (Huawei, 2024).

c **Airbnb Inc. (United States of America)** – an online marketplace for homestays and experiences. Airbnb's business model involves providing an online platform that connects hosts and guests for accommodations and travel booking. The business model is characterised by flexibility, resilience, affordability, responsible travel, and community-based tourism. During COVID-19, Airbnb faced various challenges due to travel restrictions and safety concerns. However, the company demonstrated resilience and adaptability by focusing on domestic and local travel experiences, enhanced online booking systems and cleaning protocols (Airbnb, 2020). With its continuous innovation and global reach strategies, Airbnb provides travellers personalised and unique experiences while offering hosts income opportunities.

d **Etihad Airways (United Arab Emirates)** – the second largest airline company in UAE that has utilised AI to enhance the safety management system platform. The AI enables the collection and analysis of data involving flight reports, maintenance, and training activities to accelerate and refine existing safety processes (Etihad Airways, 2023). This technology-focused approach demonstrates its commitment to maintaining industry-leading standards to continuously improve the safety and well-being of its guests and crews. Machine learning technologies such as Microsoft Azure, Google BERT, and TF-IDF are utilised to provide better analytical capabilities. Etihad also partnered with Elenium (a technology firm) to enhance its AI-enabled booking system at Abu Dhabi International Airport (Etihad Airways, 2019).

e **Arup (Thailand)** – a company diversified in various industries including rail, active travel, advanced manufacturing, hotels and leisure, education, aviation, healthcare, energy, and highways (Arup, 2024). Arup has outlined the Rail Resilience Framework focusing on three key dimensions: assets, operations, and ecosystems; leadership and strategy development; and impacts on economy and society. The framework focuses on enhancing the reliability and safety of rail services by mitigating the issues of climate change on railways, communities, and cities. Facing challenges of evolving travel behaviours and increasing urban transport congestion, Arup prioritised effective public–private collaboration and developed a whole-system resilience focusing on sustainability practices of safer, greener, and smarter rail-based transport (Sin Chew Daily, 2024).

These five case studies presented wonderful examples of how various tourism businesses have reacted, adapted, and responded to adversities and remain viable in the industry. These businesses shared some common resilience practices. First, they are committed to a continuous investment in and utilisation of technologies, which enable them to react swiftly to threats, such as global health risks. Secondly, they prioritised customer-centricity, focusing on enhancing consumer experiences. This practice helps them maintain and increase customer satisfaction and loyalty. Thirdly, they are dedicated to environmental sustainability practices, which foster good relations with other stakeholders. These businesses not only demonstrate adaptability, robustness, redundancy, and flexibility in travel systems but also serve as role models for sustainable tourism development. Stakeholders and policymakers in the travel industry could respond appropriately by implementing similar strategies in their operations.

11.6 Conclusion

This chapter defines resilience capabilities in the travel system to demonstrate how the tourism industry survives and thrives after various disruptions. The chapter also attempts to verticalise the concept of resilience in travel systems across several dimensions: adaptability, robustness, redundancy, and flexibility. By considering factors influencing resilience in travel systems, the chapter concludes that resilience occurs on four levels namely individual, organisational, community, and destination levels. This is followed by a discussion on the measurement of resilience capacities, emphasising the need for constant monitoring and evaluation of resilience strategies. Such practices help detect vulnerabilities in the travel system, implement corrective measures, and enable adaptation to dynamic challenges. The presentation of case studies illustrated successful past practices, offering innovative approaches for tourism stakeholders. As an industry vulnerable to external factors, achieving resilience capabilities in travel systems should focus on three key areas, namely technology investment, customer-centricity, and environmental sustainability practices. By focusing on these areas, stakeholders and policymakers can better understand their role in fostering resilience and ensuring the sustainability of travel systems in the face of multiple disruptions.

References

Aigbedo, H. (2021). Impact of COVID-19 on the hospitality industry: A supply chain resilience perspective. *International Journal of Hospitality Management, 98*, 103012. https://doi.org/10.1016/j.ijhm.2021.103012

Airbnb. (2020, April 27). *Airbnb's enhanced cleaning initiative for the future of travel.* https://news.airbnb.com/our-enhanced-cleaning-initiative-for-the-future-of-travel/

Arup. (2024). ARUP in Thailand. Available from https://www.arup.com/contact-us/thailand/ (Accessed 9 December 2024).

Bates, T. (1990). Entrepreneur human capital inputs and small business longevity. *The Review of Economics and Statistics*, 551–559. https://doi.org/10.2307/2109594

Bec, A., McLennan, C. L., & Moyle, B. D. (2016). Community resilience to long-term tourism decline and rejuvenation: A literature review and conceptual model. *Current Issues in Tourism, 19*(5), 431–457. https://doi.org/10.1080/13683500.2015.1083538

Bhaskara, G. I., & Filimonau, V. (2021). The COVID-19 pandemic and organisational learning for disaster planning and management: a perspective of tourism businesses from a destination prone to consecutive disasters. *Journal of Hospitality and Tourism Management, 46.* 364–375. https://doi.org/10.1016/j.jhtm.2021.01.011

Biggs, D. (2011). Understanding resilience in a vulnerable industry: the case of reef tourism in Australia. *Ecology and Society, 16*(1). https://www.jstor.org/stable/26268845

Bui, P.L., & Wickens, E. (2021). Tourism industry resilience issues in urban areas during COVID-19. *International Journal of Tourism Cities, 7*(3), 861–879. https://doi.org/10.1108/IJTC-12-2020-0289

Camilleri, M. A., & Camilleri, M. A. (2018). *The tourism industry: An overview.* Springer International Publishing. https://doi.org/10.1007/978-3-319-49849-2_1

Cheer, J., & Lew, A. (2017). *Tourism, Resilience and Sustainability.* Routledge.

Dar, S.N., Shah, S.A., & Wani, M.A. (2022). Geospatial tourist information system for promoting tourism in trans-Himalayas: A study of Leh Ladakh India. *GeoJournal, 87*(4), 3249–3263. https://doi.org/10.1007/s10708-021-10431-4

de Moraes Ramos, G., Mai, T., Daamen, W., Frejinger, E., & Hoogendoorn, S. P. (2020). Route choice behaviour and travel information in a congested network: Static and dynamic recursive models. *Transportation Research Part C: Emerging Technologies, 114*, 681–693. https://doi.org/10.1016/j.trc.2020.02.014

Dileep, M. R., & Pagliara, F. (2023). Transportation Systems and Tourism. In M. Fischer, J. C. Thill, J. van Dijk, & H. Westlund (Eds.), *Transportation Systems for Tourism* (pp. 1–25). Springer International Publishing. https://doi.org/10.1007/978-3-031-22127-9_1

Du, Y., Wang, H., Gao, Q., Pan, N., Zhao, C., & Liu, C. (2022). Resilience concepts in integrated urban transport: A comprehensive review on multi-mode framework. *Smart and Resilient Transportation, 4*(2), 105–133. https://doi/10.1108/SRT-06-2022-0013/full/html

Etihad Airways. (2019). *Etihad Airways and Elenium use ground-breaking technology to revolutionise the travel experience.* https://www.etihad.com/en/news/etihad-airways-and-elenium-use-ground-breaking-technology-to-revolutionise-the-travel-experience

Etihad Airways. (2023). *Etihad Airways pioneers cutting-edge artificial intelligence solutions to enhance safety management systems.* https://www.etihad.com/en/news/etihad-airways-pioneers-cuttingedge-artificial-intelligence-solutions-to-enhance-safety-management-systems

Feng, H., Lv, H., & Lv, Z. (2023). Resilience towarded digital twins to improve the adaptability of transportation systems. *Transportation Research Part A: Policy and Practice*, 173, 103686. https://doi.org/10.1016/j.tra.2023.103686

Ghaderi, Z., King, B., & Hall, C. M. (2022). Crisis preparedness of hospitality managers: Evidence from Malaysia. *Journal of Hospitality and Tourism Insights, 5*(2), 292–310. https://doi.org/10.1108/JHTI-10-2020-0199

Gottschalk, M., Kuntz, J. C., & Prayag, G. (2022). TouRes: Scale development and validation of a tourist resilience scale. *Tourism Management Perspectives, 44*, 101025. https://doi.org/10.1016/j.tmp.2022.101025

Hall, C. M., Prayag, G., & Amore, A. (2018). *Tourism and Resilience: Individual, Organisational and Destination Perspectives*. Channel View.

Harms, P. D., Brady, L., Wood, D., & Silard, A. (2018). Resilience and well-being. *Handbook of Well-being*, 1–12.

Harrington, R. D. (2021). Natural disasters, terrorism, and civil unrest: Crises that disrupt the tourism and travel industry-a brief overview. *Worldwide Hospitality and Tourism Themes, 13*(3), 392–396. https://doi.org/10.1108/WHATT-01-2021-0008

Hartwig, A., Clarke, S., Johnson, S., & Willis, S. (2020). Workplace team resilience: A systematic review and conceptual development. *Organizational Psychology Review, 10*(3–4), 169–200. https://doi.org/10.1177/20413866209194

Hassler, U., & Kohler, N. (2014). Resilience in the built environment. *Building Research and Information, 42*(2), 119–129. https://doi.org/10.1080/09613218.2014.873593

Huawei. (2024). *Propel airport digital transformation*. https://e.huawei.com/en/industries/aviation

Isichei, V. (2022). Hospitality, Tourism and Vulnerability. In K. Ogunyemi, E. Okoye, & O. Ogunyemi (Eds.), *Humanistic Perspectives in Hospitality and Tourism, Volume II: CSR and Person-Centred Care* (pp. 211–227). Springer International Publishing. https://doi.org/10.1007/978-3-030-95585-4_11

Kampel, K. (2022). Tourism recovery and resilience in Commonwealth Small States: Driving circular economy pathways post-COVID-19. In J. Freedman (Eds.), *IISD trade and sustainability review* (pp. 13–21). IISD.

Kaur, A., & Noonwal, V. (2023). Tourism crisis and management strategies. In A. Gupta, S. Gupta, & J. Kumar (Eds.), *Managing and strategising global business in crisis* (pp. 146–156). Routledge.

Kaushal, V., & Srivastava, S. (2021). Hospitality and tourism industry amid COVID-19 pandemic: Perspectives on challenges and learnings from India. *International Journal of Hospitality Management, 92*, 102707. https://doi.org/10.1016/j.ijhm.2020.102707

Marcouiller, D. W. (2023). Tourism: Vulnerability and resiliency in rural regions. In H. Mair (Eds.), *Handbook on tourism and rural community development* (pp. 194–204). Edward Elgar Publishing.

Masik, G. (2022). The concept of resilience: Dimensions, properties of resilient systems and spatial scales of resilience. *Geographia Polonica, 95*(4), 295–310. https://rcin.org.pl/igipz/publication/273538

Mattsson, L.G., & Jenelius, E. (2015). Vulnerability and resilience of transport systems–A discussion of recent research. *Transportation Research Part A: Policy and Practice, 81*, 16–34. https://doi.org/10.1016/j.tra.2015.06.002

Mitchell, A. (2013). *Risk and resilience: From good idea to good practice*. OECD Publishing, https://www.oecd-ilibrary.org/docserver/5k3ttg4cxcbp-en.pdf?expires=1720685719&id=id&accname=guest&checksum=12EAF3E811D7B6B7B0674B7F04F3A718

Nguyen, Q., Kuntz, J. R., Näswall, K., & Malinen, S. (2016). Employee resilience and leadership styles: The moderating role of proactive personality and optimism. *New Zealand Journal of Psychology, 45*(2), 13–21.

Nour El-Din, E., El-Halim, A., Soliman, K., & Ezzat, M. (2023). Evaluating the marketing efforts for attracting the Chinese tourists to Egypt. *International Journal of Tourism and Hospitality Management, 6*(1), 24–48. https://doi.org/10.21608/ijthm.2023.300836

OECD. (2022). *OECD Tourism Trends and Policies 2022*. OECD Publishing, https://doi.org/10.1787/a8dd3019-en

Osorio, P., Cadarso, M.Á., Tobarra, M.Á., & García-Alaminos, Á. (2023). Carbon footprint of tourism in Spain: COVID-19 impact and a look forward to recovery. *Structural Change and Economic Dynamics, 65*, 303–318. https://doi.org/10.1016/j.strueco.2023.03.003

Panzeri, A., Bertamini, M., Butter, S., Levita, L., Gibson-Miller, J., Vidotto, G., Bentall, R.P. & Bennett, K.M. (2021). Factors impacting resilience as a result of exposure to COVID-19: The ecological resilience model. *PLoS One, 16*(8), e0256041. https://doi.org/10.1371/journal.pone.0256041

Pennington-Gray, L., & Basurto-Cedeno, E. (2023). Integrated stakeholder – Centered tourism crisis. *Frontiers in Sustainable Tourism, 2*, 1–22. https://doi.org/10.3389/frsut.2023.1209325

Pocinho, M., Garcês, S., & de Jesus, S.N. (2022). Wellbeing and resilience in tourism: A systematic literature review during COVID-19. *Frontiers in Psychology.* https://doi.org/10.3389/fpsyg.2021.748947

Prasanna, R. P. I. R., Jayasundara, J. M. S. B., Naradda Gamage, S. K., Ekanayake, E. M. S., Rajapakshe, P. S. K., & Abeyrathne, G. A. K. N. J. (2019). Sustainability of SMEs in the competition: A systemic review on technological challenges and SME performance. *Journal of Open Innovation: Technology, Market, and Complexity, 5*(4), https://doi.org/10.3390/joitmc5040100

Prayag, G. (2018). Symbiotic relationship or not? Understanding resilience and crisis management in tourism. *Tourism Management Perspectives, 25*, 133–135. https://doi.org/10.1016/j.tmp.2017.11.012

Prayag, G. (2023). Tourism resilience in the 'new normal': Beyond jingle and jangle fallacies? *Journal of Hospitality and Tourism Management, 54*, 513–520. https://doi.org/10.1016/j.jhtm.2023.02.006

Prayag, G., Jiang, Y., Chowdhury, M., Hossain, M. I., & Akter, N. (2024a). Building dynamic capabilities and organizational resilience in tourism firms during COVID-19: A staged approach. *Journal of Travel Research, 63*(3), 713–740. https://doi.org/10.1177/00472875231164976

Prayag, G., Muskat, B., & Dassanayake, C. (2024b). Leading for resilience: Fostering employee and organizational resilience in tourism firms. *Journal of Travel Research, 63*(3), 659–680. https://doi.org/10.1177/00472875231164984

Prayag, G., Spector, S., Orchiston, C., & Chowdhury, M. (2020). Psychological resilience, organizational resilience and life satisfaction in tourism firms: Insights from the Canterbury earthquakes. *Current Issues in Tourism, 23*(10), 1216–1233. https://doi.org/10.1080/13683500.2019.1607832

Scott, N., & Laws, E. (2006). Tourism crises and disasters: Enhancing understanding of system effects. *Journal of Travel & Tourism Marketing, 19*(2–3), 149–158. https://doi.org/10.1300/J073v19n02_12

Sharma, N., Tabandeh, A., & Gardoni, P. (2018). Resilience analysis: A mathematical formulation to model resilience of engineering systems. *Sustainable and Resilient Infrastructure, 3*(2), 49–67. https://doi.org/10.1080/23789689.2017.1345257

SIA. (2022). Singapore Airlines and Tata Sons to merge Air India and Vistara, creating India's leading airline group. https://www.singaporeair.com/ko_KR/kr/media-centre/press-release/article/?q=en_UK/2022/October-December/ne0922-221129#:~:text=29%20November%202022%20%2D%20Singapore%20Airlines,as%20part%20of%20the%20transaction

SIA. (2023). Garuda Indonesia and Singapore Airlines propose joint venture agreement to deepen commercial partnership. https://www.garuda-indonesia.com/th/th/news-and-events/garuda-indonesia-and-singapore-airlines-propose-joint-venture-ag

Sin Chew Daily. (2024). Arup paves the way for resilient railways at Asia Pacific Rail. https://www.sinchew.com.my/news/20240531/mysinchew/5648985

Sinclair, M. T. (1998). Tourism and economic development: A survey. *The Journal of Development Studies, 34*(5), 1–51. https://doi.org/10.1080/00220389808422535

Tian, B., Stoffelen, A., & Vanclay, F. (2023). Understanding resilience in ethnic tourism communities: the experiences of Miao villages in Hunan Province, China. *Journal of Sustainable Tourism*, 1–20. https://doi.org/10.1080/14616688.2021.1938657

Tukamuhabwa, B. R., Stevenson, M., Busby, J., & Zorzini, M. (2015). Supply chain resilience: Definition, review and theoretical foundations for further study. *International Journal of Production Research, 53*(18), 5592–5623. https://doi.org/10.1080/00207543.2015.1037934

UNWTO (2020). *Tourism and COVID-19 – unprecedented economic impacts.* https://www.unwto.org/tourism-and-covid-19-unprecedented-economic-impacts

Wieczorek-Kosmala, M. (2022). A study of the tourism industry's cash-driven resilience capabilities for responding to the COVID-19 shock. *Tourism Management, 88*, 104396. https://doi.org/10.1016/j.tourman.2021.104396

Wong, S. F., & Kler, B. K. (2022). Collaborative destination marketing. In D. Buhalis (Eds.), *Encyclopedia of tourism management and marketing* (pp. 537–539). Edward Elgar Publishing.

Xiong, C., Khan, A., Bibi, S., Hayat, H., & Jiang, S. (2023). Tourism subindustry level environmental impacts in the US. *Current Issues in Tourism, 26*(6), 903–921. https://doi.org/10.1080/13683500.2022.2043835

Yang, E., Kim, J., Pennington-Gray, L., & Ash, K. (2021). Does tourism matter in measuring community resilience? *Annals of Tourism Research, 89*, 103222. https://doi.org/10.1016/j.annals.2021.103222

12 How DMOs Responded during Covid 19 Lockdowns and the Implications on Stakeholder Engagement

Andrew Higgins and Catherine McGuinn

12.1 Introduction

The purpose of this study is to investigate the implications of stakeholder engagement of the response of Destination Management Organisations (DMOs) during Covid 19 lockdowns. During the Covid 19 pandemic, many governments around the world implemented lockdown strategies, which had a direct impact on the tourism industry. While many industries were able to shift to remote working and continue operating in some capacity, the tourism sector came to an almost complete standstill due to the restrictions on movement of people being implemented during the Covid 19 lockdowns. As businesses in the tourism sector were forced to close, DMOs were able to continue to function in some capacity due to their ability to adapt to remote working. This chapter investigates the research question: What are the implications on stakeholder engagement of DMO's responses during Covid 19 lockdowns? This is of particular interest to practitioners and destination managers who could benefit from further stakeholder engagement to assist in the development of sustainable destinations that can withstand future disruptions.

12.2 Literature Review

12.2.1 The Role of the Destination Management Organisation (DMO)

The DMO plays a pivotal role in the promotion and development of the destination (Zehrer & Hallmann, 2015). For destinations where the DMO is the primary tourism organisation, they are assisted by public and private initiatives (Dredge, 2006). The DMO is considered responsible for the development of policy and marketing strategy for the destination (Zehrer & Hallmann, 2015). Due to many hospitality businesses having to close during Covid 19 lockdowns, DMOs had to pivot and adapt to implement new strategies in a scenario that none expected.

The level of disruption caused by the Covid 19 lockdowns forced destinations to respond in a way that would allow them to navigate the crisis. Opportunities would emerge for DMOs and stakeholders through new forms

DOI: 10.4324/9781032720555-13

of collaboration and stakeholder engagement through innovative technology. Collaborative initiatives can best achieve their objectives if they work together within a formal structure (Pearce, 1989). Within a collaborative initiative to achieve the required level of performance for success, it is the role of the DMO to manage stakeholder engagements (D'Angella & Go, 2009). For a destination collaboration project to succeed, there must be a strong DMO who acts to unify all stakeholders of the project and ensure stakeholders do not split from the project to form a separate group (Fyall & Garrod, 2005).

12.2.2 Stakeholder Engagement

It is widely agreed that a destination consists of many attractions, amenities, services and products, and therefore a destination encompasses many stakeholders. The tourism industry consists of a mixture of stakeholders that includes interrelated companies, government agencies and non-profit organisations (Morgan, 1996). A destination product is a complex mix of relationships with the involvement of many stakeholders (Sautter & Leisen, 1999). Collaboration amongst these stakeholders is required to accomplish the implementation of sustainable tourism (Vernon et al., 2005).

Within a destination, the competitive nature of businesses may be a potential stumbling block to any potential collaboration initiative (Wang et al., 2013). Benefits and drawbacks exist when taking a collaborative approach to destination branding. Fyall and Garrod (2005) refer to the advantages (risk reduction, mutual benefit exchange) and drawbacks (mistrust, interpersonal nature) of collaboration.

Collaborative initiatives will consist of both benefits and drawbacks, and stakeholders need to identify if collaboration is the way forward for sustainable regional development. When considering collaboration as a strategy, there are several factors that will ensure its success, but one of the most important factors is that the strategy is well-planned and of mutual benefit to all stakeholders (Aaker, 2005). Ultimately, what may be required for sustained success is leadership (Zehrer et al., 2013). A strong, passionate, energetic and committed leader is needed to guide the initiative; otherwise, the process will almost certainly fail (Baker, 2007).

There must first be motivation to collaborate amongst stakeholders (Palmer & Bejou, 1995), and planners should initially seek out stakeholders who have a willingness to adopt a collaborative strategy (Sautter & Leisen, 1999). Collaborative initiatives depend on a certain level of trust and willingness to work together amongst stakeholders (Wang & Xiang, 2007). Reluctance from businesses to collaborate with a potential rival competitor may lead to trust issues during the collaboration process (Wang et al., 2013; Naipaul et al., 2009; Sharfman et al., 1991). Previous authors have highlighted the importance of trust amongst stakeholders as a key factor for collaborative projects (Wang, et al., 2013; Bramwell & Sharman, 1999; Jamal & Getz, 1995). It may take some time for stakeholders within the collaborative arrangement to gain the trust of other members that they view as direct competitors (Wang et al., 2013; Vangen & Huxham, 2003).

It is important to engage with as many stakeholders as possible that may be affected by any proposed tourism development (Jamal & Getz, 1995). Lally et al. (2015) believe structure, membership, activities and benefits, and challenges can affect stakeholder engagement.

Palmer and Bejou (1995) assert stakeholders must recognise the potential benefits of collaboration. Once a certain level of trust has been established between stakeholders, the perceived benefits of collaboration become more prominent within stakeholder's own assessments. As a result of gained trust, there then emerges a great willingness to collaborate on a long-term basis (D' Angella & Go, 2009).

The interactions between many partners and stakeholders in a network with customers have become important (Baron et al., 2010). Cooperation between stakeholders is a key factor of a network (Bruhn, 2003). Line and Wang (2017) have introduced their framework for a multi-stakeholder market approach to destination marketing (DM) that is representative of the growing research in this area. The conceptual framework identifies key stakeholders and documents the process.

12.3 Methodology

A thorough review of research methodologies within tourism research evidently indicates that the use of qualitative research has been applied to stakeholder engagement and collaboration research (Lally et al., 2015; Wang et al., 2013).

Primary research to explore the influences of DMO's role in stakeholder engagement will inform the study. As this research is exploratory, a qualitative approach is appropriate. In-depth semi-structured interviews completed with 32 DMO managers from destinations across Ireland, Scotland, Austria, Switzerland, Italy and Australia will provide empirical evidence. Of the 32 in-depth interviews 23 are from Ireland and 9 are international. DMOs were identified as referred to in the literature (Zehrer & Hallmann, 2015), and participants were determined based on their relevance and roles in their DMO.

A combination of Judgement and Snowball sampling was utilised during the in-depth interview process. The in-depth interviews were conducted in a professional location and at a time that was best suited for both the researcher and the interviewee. The interviews were conducted face-to-face, by telephone or online video conferencing between April 2019 and September 2021. Each interview lasted between 30 and 60 minutes with some lasting 90 minutes.

The in-depth interviews were recorded and later transcribed between October 2021 and June 2022 at the research centre, ATU Sligo. A thematic analysis was conducted to identify patterns or themes within the recorded data. The use of the software package NVivo was applied between September 2022 and February 2023 to analyse the data. Manual data processing was administered between March and August 2023.

From an ethical consideration, permission was sought from all respondents taking part in interviews. When conducting face-to-face interviews, if the respondents did not want to answer a particular question, then their decision was respected. The study adhered to ATU research ethics guidelines. Participants were kept anonymous and confidential.

12.4 Findings

The findings indicate that there is a new innovative approach for collaboration between DMOs and stakeholders within a destination. The study helps inform academic researchers on a new approach to stakeholder engagement and the development of a new framework for stakeholder engagement. New strategies and policies have been identified that are key to building long-term resilience capacities in tourism networks.

Covid 19 caused many disruptions to businesses and people's lives for more than two years. These disruptions throughout the tourism industry impacted many businesses resulting in them closing for several months. This research aims to study the impact that these closures had on destination stakeholder engagement and the responses by DMOs to these disruptions. Participants were asked about their experiences of stakeholder engagement during and after Covid 19. The interview data indicates six emerging themes as indicated in Figure 12.1.

These six emerging themes include Stakeholder Engagement, Destination Assessment, Destination Recovery, Tourist Affect, Emerging Opportunities and Crisis Experience. The emerging themes will now be explored in more detail.

12.4.1 Stakeholder Engagement

Destination brand managers spoke about how they maintained stakeholder engagement throughout Covid 19. Three emerging sub-themes in stakeholder engagement consisted of Communication, Online Meetings and Increased Stakeholder Engagement (Figure 12.2).

Several managers explained how they maintained stakeholder engagement through communication and making sure that all stakeholders remained informed. Due to social distancing rules that were in place during this time, organisations were unable to hold their normal face-to-face meetings and with the emergence and rise in popularity of virtual meetings for professional work organisations utilised the technology to maintain stakeholder engagement by switching to virtual meetings. This allowed the organisations to maintain stakeholder engagement throughout the Covid 19 lockdowns. As a result of the closure of businesses due to lockdowns, business owners now found them-selves with a lot more time available to them than usual. Along with the rise in popularity and familiarity of virtual meeting technology, it created the perfect opportunity for tourism businesses to engage a lot more with destination brand managers' attempts for stakeholder engagement. Many destination brand

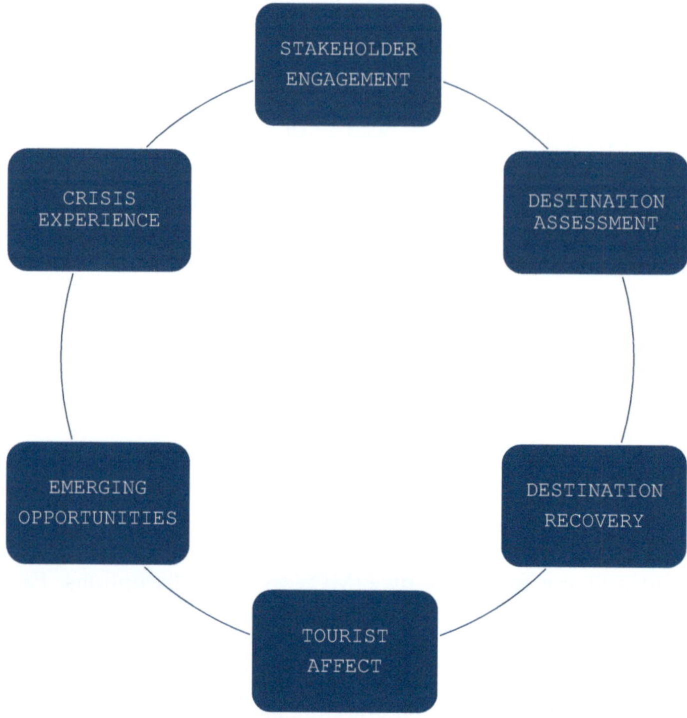

Figure 12.1 Destinations crisis response.

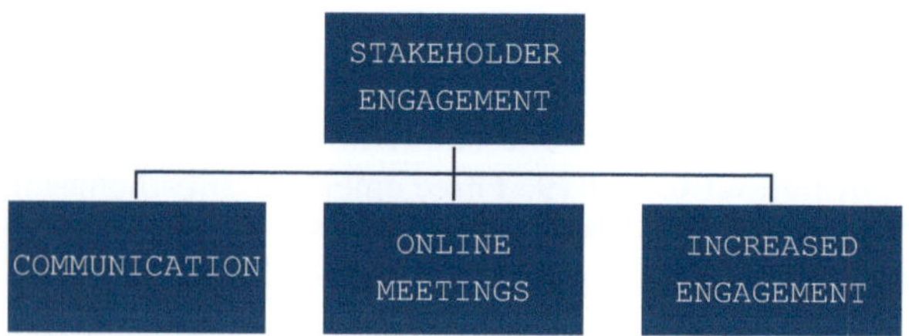

Figure 12.2 Stakeholder engagement.

managers reported an increase in the levels of tourism businesses participating in stakeholder engagement meetings through virtual meetings. One destination manager noted:

> Because of the downtime during COVID we have seen a willingness of stakeholders to have more of an input because they have more time on their hands. There is a slight shift in perception from businesses willing to collaborate and become more involved. If anything good comes out of COVID is people are now seeing that's just because there is another B&B

beside them it doesn't mean that they are in competition with each other. Everyone has their own unique selling point, and they are just starting to get their head around the idea of USP. Just because there are five down the road it doesn't mean that loads of people are staying there if they have a particular type of person staying with them.

(Respondent No. 16)

Not all destinations maintained stakeholder engagement throughout the Covid 19 lockdowns, with 4 of the 32 destinations deciding not to engage with any stakeholders during this period. The disruption of Covid 19 lockdowns caused DMOs to rethink their approach to local stakeholder engagement. The impact of not being able to hold regular face-to-face local stakeholder engagement meetings led many DMOs to pivot to online meetings. The response of DMOs to pivot to online meetings had a positive impact on local stakeholder engagement. Most managers reported an increase in the numbers of businesses participating in online stakeholder engagement meetings during Covid 19 lockdowns. These managers also stated that they intended to maintain online stakeholder engagement meetings in a blended format. Through collaboration and local stakeholder engagement, managers are building resilience in the destination to future travel disruptions. As destinations came to a stop and tourism activities were non-existent, Covid 19 lockdowns gave organisations and destination brand managers the opportunity to assess their brands.

12.4.2 *Destination Assessment*

After analyses of interview data, two sub-themes emerged from within Destination Assessment that included Destination Brand Upgrade and Brand Communication (Figure 12.3).

Destination managers talked about implementing a brand communication strategy that consisted of two phases. The first phase consisted of a Don't Visit Campaign, while the second phase comprised a Re-Opening Campaign. Managers spoke about how it was not appropriate that destinations promote themselves to attract visitors during the Covid 19 lockdowns, instead concentrating on messaging that encouraged tourists not to visit. A lot of managers

Figure 12.3 Destination assessment.

spoke about how it was the first time ever that their job consisted of having to convince tourists not to visit their destination. By conducting a Don't Visit Campaign, destinations were still able to keep their brand in people's minds. It allowed managers to prepare a Re-Opening Campaign that was ready to go when Covid 19 lockdowns were lifted. More time was available to destination brand managers to conduct a Destination Brand Upgrade. Several participants expressed how they used the opportunity of the Covid 19 lockdowns to assess what they were doing wrong and how they could improve their branding. One manager highlighted:

> After we realized it was going to be more than three weeks four weeks or five weeks we asked ourselves where can we improve so that this won't affect us if the pandemic ends next month we'll be moving so we really looked at our image our social media and how we contact people and we realized it was very poor and we had nobody doing these tasks…so we took advantage of the kindness and the fact that a lot of people were out of work can we redesign our image and pull together our social media so that they all look the same and they all had the same vibe.
>
> (Respondent No. 6)

During the disruption of Covid 19 lockdowns, DMOs had the opportunity to assess their destination. For many managers, they took the time to upgrade their destination branding. To keep their destination brand in people's minds, many DMOs initiated a brand communication strategy as part of their response to their destinations being under lockdowns. This response included Don't Visit and Re-Opening campaigns. As Covid 19 lockdowns were lifted and businesses were allowed to re-open, the focus shifted to destination recovery.

12.4.3 Destination Recovery

Destination recovery consisted of five emerging sub-themes. These sub-themes comprise Task Forces, Business Support, Staff Shortage, Temporary Business Drop-Off and Not Reliable on Tourism (Figure 12.4).

As a result of Covid 19 lockdowns, the response to the recovery of the industry included financial support to limit business closures. Tourism businesses impacted by Covid 19 lockdowns lost the ability to operate during Covid 19

Figure 12.4 Destination recovery.

lockdowns, and government financial assistance played an important role in ensuring that there were still businesses viable when re-opening. Some businesses closed permanently as the operators retired or decided to leave the industry and switch to a different industry. This resulted in a drop-off in some destinations in the amount of tourism businesses after the initial re-opening. This was also the case with regards to staff levels as businesses re-opened; they found themselves in a situation where many of their staff were no longer available to them. These staff shortages required new staff to be trained.

Recovery Task Forces were established in destinations across Ireland as a response to help strengthen collaboration and stakeholder engagement throughout the industry. Visitor Experience Development Plans (VEDPs) were in place in 18 locations across Ireland when Covid 19 lockdowns were implemented. These VEDPs, which consisted of established stakeholder networks, pivoted to become destination recovery task forces. These destination recovery task forces became the tool that would be used to help destinations in their recovery efforts and were established in destinations around the country. One manager talked in detail about the role of VEDPs:

> What we would have had is development plans two years ago are now very quite different because we will lose businesses along the way and we will have to fight hard to keep other businesses in place. Businesses that some of them offer key to the destination so we had those VDEP's networks we could pivot to destination recovery task forces or if we did not have them in place we formed them in the key destinations. We have them in 18 places around the country and they are the 18 locations which have the biggest reliance and tourism as part of the economy…They have been phenomenally successful where it worked because the key stakeholders and it sounds simple at the start it was having the coming together to share the effect. Every three weeks you come together share what it is how the restrictions or the guidelines or whatever it is are affecting your business or your sector on the ground as a representative and the fragmentation after destination so with different restrictions so meant that attractions had to be closed some restaurants were closed but if say for example attractions are closed but accommodation is opened does the destination work. No because people have a place to stay with nothing to do and the nighttime economy is closed.
>
> (Respondent No. 28)

The establishment of the recovery task forces out of the VDEP's allowed destinations to strengthen collaboration and stakeholder engagement throughout the industry. As a result, the national tourism authority intends to expand its VDEP programme across the country. Their argument is that having more destinations networked will help build resilience throughout the industry. Several participants then spoke about the effect Covid 19 has had on tourists and how their businesses interacted with tourists in the aftermath of the Covid 19 lockdowns.

Figure 12.5 Tourist affect.

12.4.4 Tourist Affect

Within Tourist Affect three sub-themes were identified consisting of Re-Opening Tourist Bounce, Popularity of Rural Areas and Behavioural Change (Figure 12.5).

With many people confined to their homes and immediate areas, there was a built-up demand for tourists wanting to travel. Many destination brand managers reported how their destinations experienced a surge in tourist numbers in the immediate aftermath of the lifting of Covid 19 lockdowns. Managers attributed the high volume of tourist numbers as a direct result of pent-up demand due to Covid 19 lockdowns. Destination managers discussed how Rural destinations became a popular choice of destination for tourists ahead of Urban destinations due to the perception that Covid 19 could spread more quickly in Urban destinations than in Rural destinations. Contributors highlighted how Rural destinations were perceived to be the safer choice by tourists. Behavioural change from tourists in the immediate aftermath of Covid 19 was identified by destination brand managers. Managers talked about changes in the expectations and attitudes of tourists, and a new way of interacting with tourists became a quick reality for many tourism businesses. One destination brand manager declared:

> People will go out to restaurants more...expectations will be higher, industry will have to offer something different, will have to do things to attract people in, can't just wait for their day trade to come in. There is not going to be as many people working in office blocks, in urban areas. Large lunch footfall won't be there so they will have to offer an experience.
>
> (Respondent No. 12)

Rural destinations saw a surge in visitor numbers, while urban destinations struggled to attract visitors to pre-Covid 19 visitor levels. Yet, several managers spoke about how the Covid 19 lockdowns gave them the opportunity to assess and identify Emerging Opportunities.

12.4.5 Emerging Opportunities

Managers highlighted development, strategy, structure and learning opportunities as four sub-themes within Emerging Opportunities (Figure 12.6).

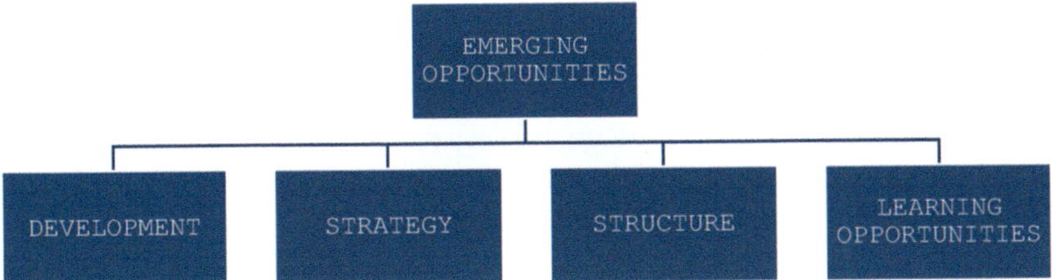

Figure 12.6 Emerging opportunities.

Managers highlighted the need to develop more rural infrastructure due to the impact of the increased popularity of rural spaces after Covid 19. Some managers identified that the structure of their organisation required a change as a response to maximise the organisation's ability to function efficiently, while other managers noted that a change in strategy would be required over the coming years as destinations recovered from the effects of Covid 19 lockdowns. The Covid 19 lockdowns gave many destination managers the potential for learning opportunities during a time of crisis. One participant stated:

> What is the nature on the ground and how can we learn from some destinations that are doing some things right and make sure that we are doing the same thing so that was it and it's genius but actually at the end of last year we sat down and looked at a kind of short term plan for each of them we got facilitated workshops in and we came up with what we were calling resilient plans. But four or five key priorities for each destination for all the stakeholders to work on together and different people took the lead on different projects as a tactical intervention over the next 12 to 18 months just to kind of make sure we get footfall, heads on beds over the next 18 months. I would say they worked well in particular, very important in a very difficult time.
>
> (Respondent No. 28)

It is evident from the research that crisis experience is an important aspect of destination management.

12.4.6 *Crisis Experience*

One destination brand manager spoke about how their destination was able to navigate Covid 19 lockdowns more successfully than other destinations due to the destination having experience dealing with crises. The destination built resilience to shocks to the tourism industry through learning from previous crisis management scenarios. The destination used its previous experiences of crisis scenarios and had a plan in place. While other destinations were panicking, this destination manager was able to implement a plan that made it easier for the destination to navigate Covid 19 lockdowns. The manager stated:

> We are lucky in a way because we are very resilient and very used to crisis and we pulled through it before because the last four years have been crisis after crisis in the destination…It started off the back of a major cyclone and then we had a series of shark attacks and a few things so that practice of having to go through crisis before made everybody stronger and they already had plans in place compared to a lot of our competing destinations were like Oh no what am I going to do because we have never been through anything like this before…We have been a couple of steps ahead through identifying what have we done before, this works, what can we do now etc. The experience of previous crisis has helped make this one a little bit better.
>
> (Respondent No. 23)

Based on these findings, the implications of the destination crisis response themes will now be discussed.

12.5 Implications

The analyses of the interview data reveal many implications for stakeholder engagement that can be implemented by destination managers during a time of crisis. These implications are listed below:

- Technology
- Recovery
- Crisis Management

The first implication is the use of technology for stakeholder engagement. Results show that during Covid 19 lockdowns, the use of technology helped increase stakeholder engagement, as stakeholders could easily attend a meeting virtually from their own homes. Managers stated that they intended to maintain virtual meetings and pursue a blended strategy of virtual and face-to-face meetings. This would allow stakeholders who previously were short on time and unwilling to attend face-to-face meetings to maintain their engagement through virtual meetings. Lally et al. (2015) refer to what affects the level of stakeholder engagement, while this research indicates that the use of technology has the potential to increase the level of stakeholder engagement.

The second implication consists of destination recovery, and within this is the most significant implication of a crisis that involves task forces. This builds on previous research on the importance of collaboration by Fyall and Garrod (2005). Destination Recovery Task Force is a network of tourism businesses collaborating in the time of crisis. Some networks or collaborative initiatives may already be established, and these can pivot to recovery task forces.

In destinations where there is no tourism network or collaborative initiative established, then a task force consisting of tourism stakeholders must be established. After the crisis is over, the task force can become the destination's tourism network, creating a resilient destination through collaboration.

The third implication involves a crisis management plan for the destination. Before Covid 19 lockdowns, not many destinations had to deal with a large crisis and found themselves in a position that they never experienced. As was shown in the data, the destination that experienced many crises in the past was better prepared to navigate the Covid 19 crisis. This was mainly due to them having a crisis management plan in place, allowing the destination to withstand the shock during Covid 19 lockdowns and be prepared for how the destination would function after the crisis. Most destinations worldwide were affected by Covid 19, so destinations have now had the experience of dealing with a crisis. It is important that destinations learn from their experiences during Covid 19 lockdowns and prepare a crisis management plan. Destinations should also consider a collaborative agreement with other destinations in the event of a crisis. By creating a network of collaborative destinations that share their crisis management experiences, destinations can become more resilient to potential shocks.

The key findings from this study show how technology can be utilised to enhance communications within stakeholder engagement. The findings highlight the importance of establishing a recovery task force to facilitate collaborations within networks. Consideration must be given by destinations to the preparation of a crisis management plan. These findings demonstrate the importance of building a network between destinations to help develop a resilient tourism industry.

The most recent strategy documents for regional tourism development in Ireland (Fáilte Ireland, 2023a, 2023b, 2023c, 2023d) across Fáilte Ireland's four regional brands establish strategic objectives in community stakeholder engagement. Fáilte Ireland is building on the recovery task forces that were established during Covid 19 lockdowns to establish Destination and Experience Development Plans (DEDPs). These DEDPs rely on collaboration between stakeholders and industry partners. By expanding DEDPs across the four regional brands of Ireland, Fáilte Ireland is strengthening local destinations' resilience against any potential future disruptions.

12.6 Conclusions

This chapter has examined how DMOs responded during Covid 19 lockdowns and the subsequent implications on stakeholder engagement. The importance of technology for maintaining and increasing stakeholder engagement during Covid 19 lockdowns has been documented in this research. Throughout the interviews, destination managers intend to utilise the use of technology through virtual meetings alongside face-to-face meetings to maintain the levels of stakeholder engagement experienced during Covid 19 lockdowns.

This represents a shift in the way DMOs now approach stakeholder engagement, from the more traditional approach to a more modern, blended approach.

Furthermore, the research reveals the significance of collaboration within a destination, and a well-networked destination is better placed to withstand potential shocks to the destination. Destinations that already had collaborative initiatives or established networks were able to pivot to recovery task forces quickly. Destinations should consider establishing collaborative initiatives or networks to better insulate themselves from potential crises.

If tourism destinations are to become more resilient to potential shocks, then they must have a crisis management plan in place. This involves all different types of potential shocks that destinations need to plan for, which includes a during and after phase of the crisis. Another way to build resilience amongst destinations is by establishing a destination collaborative initiative where destinations share their experiences of dealing with different crises. If a destination suffers a crisis that they have no experience of, they can collaborate with another destination that has experience in that specific scenario. This research has identified the impacts and responses of DMOs during Covid 19 lockdowns and has shown that destinations can build resilience through collaboration.

References

Aaker, D. (2005) *Strategic market management.* 7th ed. New Jersey: John Wiley and Sons, Inc.

Baker, B. (2007) *Destination branding for small cities: The essentials for successful place branding.* 1st ed. Oregon: Creative Leap Books.

Baron, S., Conway, T. & Warnaby, G. (2010) *Relationship marketing. A consumer approach.* California: Sage Publications. https://doi.org/10.4135/9781446251096

Bramwell, B. & Sharman, A. (1999) "Collaboration in local tourism policy making". *Annals of Tourism Research* 26(2), pp. 392–415. https://doi.org/10.1016/S0160-7383(98)00105-4

Bruhn, M. (2003) *Relationship marketing. Management of customer relationships.* Essex: Pearson Education Ltd.

D'Angella, F. & Go, F. (2009) "Tale of two cities collaborative tourism marketing: Towards a theory of destination stakeholder assessment". *Tourism Management* 30, pp. 429–440. https://doi.org/10.1016/j.tourman.2008.07.012

Dredge, D. (2006) "Policy networks and the local organisation of tourism". *Tourism Management* 27, pp. 269–280. https://doi.org/10.1016/j.tourman.2004.10.003

Fáilte Ireland (2023a) "Dublin regional tourism development strategy 23–27" https://www.failteireland.ie/FailteIreland/media/WebsiteStructure/Documents/Dublin/Dublin-Regional-Tourism-Development-Strategy.pdf

Fáilte Ireland (2023b) "Hidden heartlands regional tourism development strategy 23–27" https://www.failteireland.ie/FailteIreland/media/WebsiteStructure/Documents/Ireland's%20Hidden%20Heartlands/Ireland-s-Hidden-Heartlands-Regional-Tourism-Development-Strategy.pdf

Fáilte Ireland (2023c) "Irelands ancient east regional tourism development strategy 23–27" https://www.failteireland.ie/FailteIreland/media/WebsiteStructure/Documents/Irelands%20Ancient%20East/Ireland-s-Ancient-East-Regional-Tourism-Development-Strategy-Web.pdf

Fáilte Ireland (2023d) "Wild Atlantic way regional tourism development strategy 23–27" https://www.failteireland.ie/FailteIreland/media/WebsiteStructure/Documents/Wild%20Atlantic%20Way/Wild-Atlantic-Way-Regional-Tourism-Development-Strategy.pdf

Fyall, A. & Garrod, B. (2005) *Tourism marketing. A collaborative approach.* 1st ed. Clevedon: Channel View Publications. https://doi.org/10.21832/9781873150917

Jamal, T. & Getz, D. (1995) "Collaboration Theory and Community Tourism Planning". *Annals of Tourism Research* 22(1), pp. 186–204. https://doi.org/10.1016/0160-7383(94)00067-3

Lally, A. M., O'Donovan, I. & Quinlan, T. (2015) "Stakeholder engagement in destination management: Exploring key success factors". *11th Annual Tourism and Hospitality Research in Ireland Conference (THRIC)*, 11–12 June.

Line, N. & Wang, Y. (2017) "A multi-stakeholder market oriented approach to destination marketing". *Journal of Destination Marketing and Management* 6, pp. 84–93. https://doi.org/10.1016/j.jdmm.2016.03.003

Morgan, M. (1996) *Marketing for leisure and tourism.* 1st ed. London: Prentice Hall.

Naipaul, S., Wang, Y. & Okumus, F. (2009) "Regional destination marketing: A collaborative approach". *Journal of Travel and Tourism Marketing,* 26(5/6), pp. 462–481. https://doi.org/10.1080/10548400903162998

Palmer, A. & Bejou, D. (1995) "Tourism destination marketing alliances". *Annals of Tourism Research* 22(3), pp. 616–629. https://doi.org/10.1016/0160-7383(95)00010-4

Pearce, D. (1989) *Tourist development.* New York: John Wiley and Sons, Inc.

Sautter, E & Leisen, B. (1999) "Managing stakeholders. A tourism planning model". *Annals of Tourism Research* 26(2), pp. 312–328. https://doi.org/10.1016/S0160-7383(98)00097-8

Sharfman, M. P., Gray, B. & Yan, A. (1991) "The context of interorganizational collaboration in the garment industry: an institutional perspective". *The Journal of Applied Behavioural Science* 27(2), pp. 181–208. https://doi.org/10.1177/0021886391272003

Vangen, S. & Huxham, C. (2003) "Nurturing collaborative relations: Building trust in interorganizational collaboration". *The Journal of Applied Behavioural Science* 39(5), pp. 5–31. https://doi.org/10.1177/0021886303039001001

Vernon, J., Essex, S., Pinder, D. & Curry, K. (2005) "Collaborative policymaking local sustainable projects". *Annals of Tourism Research,* 32(2), 325–345. https://doi.org/10.1016/j.annals.2004.06.005

Wang, Y., Hutchinson, J., Okumus, F. & Naipaul, S. (2013) "Collaborative marketing in a regional destination: Evidence from Central Florida". *International Journal of Tourism Research* 15, pp. 285–297. https://doi.org/10.1002/jtr.1871

Wang, Y. & Xiang, Z. (2007) "Toward a theoretical framework of collaborative destination marketing". *Journal of Travel Research* 46, pp. 75–85. https://doi.org/10.1177/0047287507302384

Zehrer, A. & Hallmann, K. (2015) "A stakeholder perspective on policy indicators of destination competitiveness". *Journal of Destination Marketing and Management* 4, pp. 120–126. https://doi.org/10.1016/j.jdmm.2015.03.003

Zehrer, A., Raich, F., Siller, H. & Tschiderer, F. (2013) "Leadership networks in destinations". *Tourism Review* 69(1), pp. 59–73. https://doi.org/10.1108/TR-06-2013-0037

13 Tourist resilience from geography

Proposals or responses in Spanish spa tourism?

Aida Pinos-Navarrete

13.1 Introduction

The dynamics of globalisation have shaped a tourism model in which visitor flows predominantly head towards saturated spaces (overtourism), where the excess of tourism generates significant problems and challenges (Dodds & Butler, 2019; Mihalic, 2020). In the current tourism paradigm, tourists often travel due to the desire to follow travel trends set by social media, and along the way, the territory and local society become merely a backdrop without significant prominence (Toro & Pinos, 2022). Additionally, mass tourism has monopolised contemporary forms of leisure and recreation, leading to significant territorial polarisation. This scenario and its resulting negative impacts have been somewhat reinforced by the emergence of post-tourism as a new form of articulating the tourism practice (Toro & Pinos, 2022). In this context, and as a reaction to the need to alleviate undesirable impacts at all scales and dimensions, it is necessary to implement the concept of resilience in the tourism industry (Prayag, 2023).

Tourism resilience is positioned as a desirable asset. Containing growth to prevent destination saturation, through protection, adaptation, and recovery processes, would be a possible solution or mitigation to the negative effects of tourism massification (Cheer & Lew, 2017). Overtourism poses a challenge that needs to be addressed with a resilient alternative proposal for the development or subsistence of those areas facing demographic, economic, social, and environmental challenges (Fletcher et al., 2020). In contrast to the overtourism model, new leisure and recreation practices can be more resilient and sustainable if they start by relocating tourist flows and qualitatively and quantitatively controlling them towards non-saturated spaces. This circumstance becomes an opportunity for lesser-known and more selective destinations, such as rural areas with less tourist influx. With this new direction and focus, it is necessary to examine the resilience capacity of potential new tourist destinations and the alternative tourism types that could develop therein.

In this context, spa tourism, based on the use of mineral-medicinal waters, could be explored as a resilient tourism modality that generally develops in rural areas of some European countries. Thermalism is a global phenomenon

DOI: 10.4324/9781032720555-14

with a notable historical tradition. Since antiquity, Europe's healing hot waters have attracted the interest of various civilisations. This ancestral relationship between humans and mineral-medicinal waters has created a synergistic bond between the population, their territory, and their water resources. The history of how tourists have used these waters for health and well-being has shaped a genuine thermal culture that forms part of Europe's history and identity. The current state of spa tourism is the result of the interaction of different cultures that have mutually influenced each other in the use of mineral-medicinal waters at certain times in the past (Van Tubergen & Van der Linden, 2002; Alonso-Álvarez, 2012).

However, the development of this tourism modality has not been constant over time nor homogeneous in territorial terms, continually facing various challenges. Territorial differences in the thermal paradigm can be very significant (Alonso-Álvarez & Larrinaga-Rodríguez, 2015). Furthermore, alongside new demand motivations, spas exist in a dynamic space and time that involves numerous interrelationships and constant changes. All of these, due to their scale and immediacy, require rapid responses from spa tourism, and certainly of a resilient nature, as the only way to ensure its continuity, identity, and synergistic relationship with the territory in which it develops. In this scenario, the objective is to theoretically analyse whether spa tourism in Spain can contribute to the resilience of the tourism industry.

13.2 Historical development of Spanish spa tourism in the European context: Inter- and intra-regional differences

In the 21st century, spas are no longer solely synonymous with health, but also with aesthetics and, above all, leisure, highlighting their recreational component (Araujo & Fraiz, 2012). The recent evolution of the thermal sector in Europe describes a general process that could be termed "spaization," understood as the progressive disappearance of the traditional medical function of spas alongside an increase in the prominence of the recreational-preventive dimension (Pinos et al., 2021).

According to Alonso-Álvarez and Larrinaga-Rodríguez (2015), theoretical, conceptual, and historical analysis reflects that the evolution of thermalism has shaped different models of thermal tourism in European regions. However, this complex assertion is influenced by many more factors. Southern Europe (including Spain) maintains some common characteristics with certain national and local nuances. These thermal destinations are characterised by a significant uniqueness and the persistence of popular tradition, reminiscent of the therapeutic-medical use by Roman and Muslim civilisations. However, this therapeutic application is scarcely noticeable in the countries of Northern Europe, where the recreational and preventive component dominates the current offer, having been implemented several decades ago. The advent of the Grand Tour, with stops at thermal stations in these countries, coupled with the purchasing power of these British travellers, accelerated the incorporation of non-therapeutic elements in these destinations, simultaneously making them centres of economic, social, and cultural dynamism with an international projection. The lesser economic development of the south and its limited presence on the routes established by the Grand Tour explain, with few exceptions, the significant contrasts in spa tourism between Northern and Southern Europe. This economic dimension was accompanied by weak business organisation, insufficient and/or non-existent regulations, poor technological investment, and less-controlled hygiene (Alonso-Álvarez & Larrinaga-Rodríguez, 2015). With these premises, thermal culture in the northern and southern countries developed with marked differences that are evident today, with their advantages and disadvantages (Legido Soto et al., 2009; Pinos-Navarrete et al., 2022).

On the one hand, the lack of attention and appreciation for thermal tourism in the south has led to slower sector development, while allowing the conservation of its traditional uniqueness due to the prominent absence of a reconceptualisation orchestrated by supply and/or demand. On the other hand, the aforementioned reconceptualisation forms part of an adaptation of thermalism to demand, as well as a demonstration of its resilience by attempting to preserve its original concept. These conceptual differences in spa tourism, considering its geographical location and spatial characteristics, condition the resilience capacity of this tourism type in social, economic, and environmental dimensions today.

13.3 Possibilities and limitations for the resilience of thermal-tourism territories. Proposals and responses in the current Spanish context. An issue for discussion

Experts in thermal tourism highlight the need to offer more effective, eco-friendly, and sustainable treatments, coupled with authentic local experiences (Gianfaldoni et al., 2017; Stevens et al., 2018; Smith & Wallace, 2020). Thermal establishments can no longer afford to be perceived solely as elite tourism destinations but must instead serve as healthy and sustainable centres for society at large. This approach would revive the more therapeutic dimension of thermal tourism in Europe, a focus that has been overshadowed in recent decades by wellness research. Authors such as Cassens et al. (2012), who analysed health tourism opportunities in Germany, and Speier (2008), who studied the health benefits of spas in the Czech Republic, have highlighted the importance of this therapeutic aspect. In this sense, Spain's case is notably resilient due to its maintained original identity throughout its evolution.

Furthermore, the role of the local community must be conceived in a dual manner, as it is also important for the local community to feel integrated and actively involved in the resilient management of their water resource. This integration would enable the development of a thermal model beneficial for all members of the tourism system and resilient in the face of current challenges. Thus, spatial planning should incorporate spa tourism, including all territorial elements and agents, to ensure the overall system functions correctly. These issues have been directly analysed by authors such as Surdu et al. (2015), Negrea et al. (2016), Loke et al. (2018), and Szromek (2020, 2021).

This endogenous and "bottom-up" approach could significantly boost local economies (García-Altés, 2005). The economic dimension analysis already conducted by Laczó & Ács (2009) on Hungarian spas, Drăghici et al. (2016) on Slovakian spas, and Derco & Pavlisinova (2017) on Romanian spas should not overlook other dimensions of reality that cannot be quantified monetarily. Just as important as the economic development of European thermal tourist destinations is identifying the elements hindering this development, as examined by Anaya-Aguilar, Gemar & Anaya-Aguilar (2021), as well as understanding their territorial impact, as explored by Jónás-Berki et al. (2015). Spain has one of the highest numbers of spas in Europe, with a total of 114 active establishments. 80% of these are located outside urban areas, in regions with very low population densities. However, they possess exceptional potential to be exploited sustainably for tourism and to help improve the reality of rural populations. The objective should be to analyse the sector to replicate successful resilient cases in the new context of the tourism industry, as documented by Martyin (2015), with a more sustainable model aimed at local development, as Papageorgiou and Beriatos (2011) suggest, including the population's perception, as in the case of Michalkó et al. (2013).

On the other hand, the economy and management of thermal establishments must deal (resiliently manage) with the intricacies of current Western lifestyle patterns. Generally, spa users have little free time due to excessive work and desire significant results from their thermal stay in the shortest possible time. Moreover, holistic experiences are valued more than purely medical ones, encompassing healthy eating, fitness, mindfulness, and even spirituality. These changes in the motivations of thermal tourists have been recurrently addressed by the author Dryglas in her research focused on Poland (Dryglas & Różycki, 2016, 2017; Dryglas & Salamaga, 2017) and can be extrapolated, within limits, to other European countries. In this context, opening up to new client profiles in Spain should be a mandatory strategy for the benefit of the sector and the positive impacts of this openness on society. However, the strategic resilience of this type of tourism lies not only in the characteristics of the demand but also in the characteristics of the industry, as thermal tourism is a sector with great potential for counteracting seasonality, offering tourism diversification, protecting natural resources, and geographically decentralising the supply. It also provides opportunities for creating qualified and sustainable employment by proposing extended stays to ensure the effectiveness of thermal cures and having a higher average tourist expenditure compared to other forms of tourism.

The changes in these characteristics of the stay in recent decades in some European countries, which seem to promote quantity over quality in this type of tourism, should not be forgotten, especially in the current tourism context, the defining strengths of spa tourism. In localities whose economy revolves around health tourism, the territory, natural resources, heritage, population, and landscape are prominent parts of the concept that have historically contributed to the success of thermal therapies and the well-being of users and the local population. Managing these elements as integral parts of tourism activity is crucial to maintaining the relationships that thermal activity has had with its municipalities and among thermal tourists, as walks, local resources, and the landscape have created and enhanced local and regional identities (Navarro-García & Alvim-Carvalho, 2019). Spanish thermal destinations should not fall into overtourism and should continue to prioritise the conservation and offering of their cultural heritage, landscape quality, natural environments, and a comprehensive and updated conception of health (Navarro-García & Alvim-Carvalho, 2019). Additionally, preserving the concept and environment is essential, as noted by Valeriani et al. (2018), since the mineromedicinal waters of the spas are a natural resource that must be preserved to maintain their therapeutic characteristics and quality. Concurrently, the environment surrounding the thermal cure has potential therapeutic effects and enhances psychophysical health (Antonelli et al., 2021). Therefore, the environment should be considered another resource of tourism activity that cannot be altered and that shapes the therapeutic landscape.

This added value, the symbiotic relationship with its surroundings, is what gives spa tourism its uniqueness and capacity for distinction in the global thermal market. Tradition, rootedness, and the therapeutic uniqueness of the Spanish case are key to positioning itself as a health brand against emerging competitors, such as spas. Studies by leading authors in spa tourism, such as Pforr and Locher (2012) and Alén et al. (2014), have highlighted the importance of creating an offering that allows spas to distinguish themselves and offer a unique product in the market. Some studies, such as those by Lebe (2006), Smith (2015), and Sziva et al. (2017), address the image of a health tourism brand in the Balkans as a potential future development strategy, although it is more of a strategy than a current reality. Likewise, Fontanari and Kern (2003) have also analysed the context of opportunity that positioning as a health destination brand represents for increasing attractiveness.

The uniqueness of the Spanish sector, characterised by being a non-mass and proximity tourism, attracts a more conscientious tourist. This issue has been addressed by Alén et al. (2007) for Spain, concluding that positive perceived quality increases word-of-mouth communication and purchase intentions while decreasing price sensitivity. In the current situation, despite new challenges, thermal tourism by definition is a type of tourism that aligns with the new strategic lines of the sector in favour of sustainability, degrowth, and resilience. Managing spas in this direction, away from mass and over-consumed tourism, will allow spa tourism to emerge and position itself sustainably and resiliently, both for its development and to ensure its longevity, with the commitment to sustainable growth in the territories where it is located.

In addition, it is undeniable that in the face of social and circumstantial changes, thermal establishments must offer a rapid response and adapt to be as resilient as possible. The guarantee of survival in the current tourism and socio-health context (post-pandemic) requires a reinvention and offering of Spanish spas in line with the thermal community and the new scenarios being presented.

This new context provides a holistic view of the reality of demand in spa tourism and the possible socio-health changes post-COVID-19 pandemic. It confirms a shift in the needs and preferences of spa tourism clients compared to previous decades. The evolution of spa tourism towards wellness (Smith & Puczkó, 2015; Kasagranda & Gurňák, 2017) may deviate from its general trend following the recent health crisis. The return of thermalism to its more original conception, in its more medical-health dimension (Freire, 2020), has already been identified as an opportunity by Pinos and Shaw (2020) for the Spanish case. The sector, therefore, finds itself in a possible new scenario, different from what has been indicated in studies such as Henn et al. (2008), Smith and Puczkó (2010), or Pinos et al. (2021). Consequently, it is time for thermalism and its clients to readapt, as has occurred in other historical periods of reform documented by authors such as Pforr and Locher (2012). We are facing new demands somewhat different from those previously experienced and analysed by Rancic et al. (2014) or Dryglas and Różycki (2017). Additionally, the way and importance in which clients perceive the establishment, segmented into three clusters by Kamenidou et al. (2014), may change significantly, now enhancing the first option, "clients with high health demands," which implies a demand for higher quality and incentivises the resilience of the Spanish offering.

In this context, new opportunities and challenges may arise for the sector in Spain. Simultaneously, it becomes essential to reconsider the benefits for local and social development derived from recent conceptual changes in spa tourism. The trend followed by some spas in European countries, such as the United Kingdom or Germany, towards a reorientation of their functions more linked to aesthetics, prevention, and recreational use—in essence, towards wellness—now results in a certain incapacity to offer a rapid medical response, limiting their resilient reaction capability compared to other countries like Spain, which are better positioned to reinvent themselves according to the needs of the supply, demand, and the situation of the global tourism industry.

Therefore, practical recommendations for the spa industry in Spain (and the EU) are fundamentally based on the need to manage the establishment's tourism activity sustainably, for both internal benefits and those of the surrounding environment. Moreover, it is essential to offer a holistic experience that is flexible in responding to demand, while being rigorous in maintaining the uniqueness and distinction of its offerings as a hallmark of quality. In this sense, it is essential to adopt strategic resilience in management to ensure and strengthen economic, social, and environmental sustainability.

13.4 Conclusions

The dynamics of globalisation and the pandemic have had a significant impact on the tourism industry. The resilience of tourist destinations, including spas, has depended on their ability to adapt to new conditions, implement reorientation measures, diversify their offerings, and respond to the changing expectations of users in the evolution of spa tourism.

Thermal tourism in Spain plays a fundamental role in the resilience of the tourism industry against challenges such as overtourism and environmental, economic, and social problems. By diversifying the tourism offer, spas attract a segment of visitors interested in wellness and health, reducing the pressure on other conventional and highly crowded destinations, where overtourism predominates. This diversification not only decreases the impact of mass tourism in certain areas but also expands economic opportunities across the country, especially in the less-visited regions where most spas are located. Additionally, this diversification can help attract different segments of tourists and increase their resilience by adapting to more heterogeneous demand profiles.

Spas are also active practically all year round, contributing to the destabilisation of tourism in the high season. By attracting tourists outside of that period, it avoids the overload during the summer months and holidays, allowing for a more even distribution of visitors and their impacts on the territory. This is crucial to mitigate the environmental, economic, and social pressure on destinations and promote territorial resilience. The trend towards overcoming seasonality as one of its main characteristics generates a more stable base of occupancy and wealth, as well as business profitability, resulting in resilience in the social and economic dimensions. Spa tourism also tends to attract tourists with a higher spending capacity, interested in high-quality experiences in less crowded destinations. These visitors seek wellness services, therapies, and personal care, generating higher, more stable, and more sustainable income for the local populations and economies where they are located. The creation of stable jobs throughout almost the entire year would necessarily imply, indirectly, the development of new complementary and economically resilient activities.

Finally, the location of many Spanish spas in natural environments promotes the conservation of these spaces and their natural, heritage, and cultural resources. By focusing on the use of thermal waters with therapeutic properties, spas can significantly contribute to the environmental sustainability of tourism in Spain. This natural attraction can be a resilient factor, as the demand for wellness and health experiences can persist even in times of economic, social, and environmental challenges.

In summary, thermal tourism in Spain, based on the use of mineral-medicinal waters, could be a sustainable and resilient form of tourism that meets the needs of rural areas and the tourism industry in general. This aligns with the definition of the degrowth model that should be addressed for the benefit of resilience applied to the tourism industry.

References

Alén, E., De Carlos, P., & Domínguez, T. (2014). An analysis of differentiation strategies for Galician thermal centres. *Current Issues in Tourism*, 17(6), 499–517. https://doi.org/10.1080/13683500.2012.733357

Alén, E., Rodríguez, L., & Fraiz, J. A. (2007). Assessing tourist behavioral intentions through perceived service quality and customer satisfaction. *Journal of Business Research*, 60(2), 153–160. https://doi.org/10.1016/j.jbusres.2006.10.014

Alonso-Álvarez, L. (2012). The value of water: The origins and expansion of thermal tourism in Spain, 1750–2010. *Journal of Tourism History*, 4(1), 15–34. https://doi.org/10.1080/1755182X.2012.671373

Alonso-Álvarez, L., & Larrinaga-Rodríguez, C. (2015). Health tourism and welfare in Southern Europe. *Water and Landscape* (6), 8–11. https://doi.org/10.17561/at.v0i6

Anaya-Aguilar, R., Gemar, G., & Anaya-Aguilar, C. (2021). Challenges of spa tourism in Andalusia: Experts' proposed solutions. *International Journal of Environmental Research and Public Health*, 18(4), 1–17. https://doi.org/10.3390/ijerph18041829

Antonelli, M., Donelli, D., Veronesi, L., Vitale, M., & Pasquarella, C. (2021). Clinical efficacy of medical hydrology: An umbrella review. *International Journal of Biometeorology*, 65(10), 1597–1614. https://doi.org/10.1007/s00484-021-02133-w

Araujo, N., & Fraiz, J. A. (2012). Los establecimientos termales como atractivo turístico del siglo XXI y dinamizadores del desarrollo local. *Revista de Investigación en Turismo y Desarrollo Local*, 5(12), 15.

Cassens, M., Hörmann, G., Tarnai, C., Stosiek, N., & Meyer, W. (2012). Trend Gesundheitstourismus: Steigende Bedeutung des touristischen Settings für Gesundheitsförderung und medizinische Prävention. *Pravention Und Gesundheitsforderung*, 7(1), 24–29. https://doi.org/10.1007/s11553-011-0313-2

Cheer, J. M., & Lew, A. A. (2017). Understanding tourism resilience: Adapting to social, political, and economic change. In *Tourism, resilience and sustainability* (pp. 3–17). Routledge. https://doi.org/10.4324/9781315464053-1

Derco, J., & Pavlisinova, D. (2017). Financial position of medical spas–The case of Slovakia. *Tourism Economics*, 23(4), 867–873. https://doi.org/10.5367/te.2016.0553

Dodds, R., & Butler, R. (2019). The phenomena of overtourism: A review. *International Journal of Tourism Cities*, 5(4), 519–528. https://doi.org/10.1108/IJTC-06-2019-0090

Drăghici, C. C., Diaconu, D., Teodorescu, C., Pintilii, R. D., & Ciobotaru, A. M. (2016). Health tourism contribution to the structural dynamics of the territorial systems with tourism functionality. *Procedia Environmental Sciences*, 32, 386–393. https://doi.org/10.1016/j.proenv.2016.03.044

Dryglas, D., & Różycki, P. (2016). European spa resorts in the perception of non-commercial and commercial patients and tourists: The case study of Poland. *E-Review of Tourism Research*, 13(1–2), 382–400.

Dryglas, D., & Różycki, P. (2017). Profile of tourists visiting European spa resorts: A case study of Poland. *Journal of Policy Research in Tourism, Leisure and Events*, 9(3), 298–317. https://doi.org/10.1080/19407963.2017.1297311

Dryglas, D., & Salamaga, M. (2017). Applying destination attribute segmentation to health tourists: A case study of Polish spa resorts. *Journal of Travel and Tourism Marketing*, 34(4), 503–514. https://doi.org/10.1080/10548408.2016.1193102

Fletcher, R., Mas, I. M., Blanco-Romero, A., & Blázquez-Salom, M. (2020). Tourism and degrowth: An emerging agenda for research and praxis. In *Tourism and Degrowth* (pp. 1–19). Routledge. https://doi.org/10.4324/9781003017257-1

Fontanari, M., & Kern, A. (2003). The "comparative analysis of spas" – An instrument for the re-positioning of spas in the context of competition in spa and health tourism. *Tourism Review*, 58(3), 20–28. https://doi.org/10.1108/eb058413

Freire Magariños, A. (14th April 2020). *Utilidades preventivas y terapéuticas de la terapia termal respiratoria, riesgos y beneficios*. Webinar – Termatalia.

García-Altés, A. (2005). The development of health tourism services. *Annals of Tourism Research*, 32(1), 262–266. https://doi.org/10.1016/j.annals.2004.05.007

Gianfaldoni, S., Tchernev, G., Wollina, U., Roccia, M. G., Fioranelli, M., Gianfaldoni, R., & Lotti, T. (2017). History of the baths and thermal medicine. *Open Access Macedonian Journal of Medical Sciences*, 5, 566–568. https://doi.org/10.3889/oamjms.2017.126

Henn Bonfada, M. R., Branco Bonfada, P. L., Gonçalves Gandara, J. M., & Fraiz Brea, J. A. (2008). Turismo termal: cambios conceptuales y mercadológicos de los balnearios en España. *Turismo-visão e ação*, 10(3), 415–434.

Jónás-Berki, M., Csapó, J., Pálfi, A., & Aubert, A. (2015). A market and spatial perspective of health tourism destinations: The Hungarian experience. *International Journal of Tourism Research*, 17(6), 602–612. https://doi.org/10.1002/jtr.2027

Kamenidou, I. C., Mamalis, S. A., Priporas, C.-V., & Kokkinis, G. F. (2014). Segmenting customers based on perceived importance of wellness facilities. *Procedia Economics and Finance*, 9, 417–424. https://doi.org/10.1016/s2212-5671(14)00043-4

Kasagranda, A., & Gurňák, D. (2017). Spa and wellness tourism in Slovakia (A geographical analysis). *Czech Journal of Tourism*, 6(1), 27–53. https://doi.org/10.1515/cjot-2017–0002

Laczó, T., & Ács, P. (2009). Spatial characteristics of the Hungarian wellness market's demand and supply relations. *World Leisure Journal*, 51(3), 197–210. https://doi.org/10.1080/04419057.2009.9728272

Lebe, S. S. (2006). European spa world: Chances for the project's sustainability through application of knowledge management. *Journal of Quality Assurance in Hospitality & Tourism*, 7(1–2), 137–146. https://doi.org/10.1300/J162v07n01_08

Legido Soto, J. L., Mourelle Mosqueira, M. L., Medina Filgueira, C. M., Gómez Pérez, C. & Meijide Faílde, R. M. (2009). *Termalismo: Aspectos generales*. Universidad de Vigo.

Loke, Z., Kovács, E., & Bacsi, Z. (2018). Assessment of service quality and consumer satisfaction in a Hungarian spa. *Deturope*, 10(2), 124–146. https://doi.org/10.32725/det.2018.017

Martyin, Z. (2015). A dynamically developing Hungarian spa town: Mórahalom. *European Journal of Geography*, 6(1), 37–50. https://doi.org/10.1016/j.jhg.2015.05.003

Michalkó, G., Bakucz, M., & Rátz, T. (2013). The relationship between tourism and residents' quality of life: A case study of Harkány, Hungary. *European Journal of Tourism Research*, 6(2), 154–169. https://doi.org/10.54055/ejtr.v6i2.129

Mihalic, T. (2020). Conceptualising overtourism: A sustainability approach. *Annals of Tourism Research*, 84, 103025. https://doi.org/10.1016/j.annals.2020.103025

Navarro-García, J. R., & Alvim-Carvalho, F. (2019). *Paisaje y salud: enfoques y perspectivas del termalismo en España*. Jaén: Universidad de Jaén.

Negrea, A., Cosma, M. R., & Popescu, M. L. (2016). Sustainable development of spa tourism in Romania. *Quality – Access to Success*, 17, 412–414.

Papageorgiou, M., & Beriatos, E. (2011). Spatial planning and development in tourist destinations: A survey in a Greek spa town. *International Journal of Sustainable Development and Planning*, 6(1), 34–48. https://doi.org/10.2495/SDP-V6-N1-34-48

Pforr, C., & Locher, C. (2012). The German spa and health resort industry in the light of health care system reforms. *Journal of Travel and Tourism Marketing*, 29(3), 298–312. https://doi.org/10.1080/10548408.2012.666175

Pinos, A., Sánchez, L. M., & Maroto, J. C. (2021). El turismo de balneario en Europa Occidental: Reconceptualización y nuevas funciones territoriales en una perspectiva comparada. *Boletín de la Asociación De Geógrafos Españoles* (88). https://doi.org/10.21138/bage.3061

Pinos, A., & Shaw, G. (2020). Spa tourism opportunities as strategic sector in aiding recovery from COVID-19: The Spanish model. *Tourism and Hospitality Research*, 21(2), 245–250. https://doi.org/10.1177/1467358420970626

Pinos-Navarrete, A., Abarca-Álvarez, F. J., & Maroto-Martos, J. C. (2022). Perceptions and profiles of young people regarding spa tourism: A comparative study of students from Granada and Aachen Universities. *International Journal of Environmental Research and Public Health*, 19(5), 2580. https://doi.org/10.3390/ijerph19052580.

Prayag, G. (2023). Tourism resilience in the 'new normal': Beyond jingle and jangle fallacies? *Journal of Hospitality and Tourism Management*, 54, 513–520. https://doi.org/10.1016/j.jhtm.2023.02.006

Rancic, M., Pavic, L., & Mijatov, M. (2014). Wellness centers in Slovenia: Tourists' profiles and motivational factors. *Turizam*, 18(2), 72–83. https://doi.org/10.5937/turizam1402072r

Smith, M. (2015). Baltic health tourism: Uniqueness and commonalities. *Scandinavian Journal of Hospitality and Tourism*, 15(4), 357–379. https://doi.org/10.1080/15022250.2015.1024819

Smith, M., & Puczkó, L. (2010). Taking your life into your own hands? New trends in European health tourism. *Tourism Recreation Research*, 35(2), 161–172. https://doi.org/10.1080/02508281.2010.11081631

Smith, M., & Puczkó, L. (2015). More than a special interest: Defining and determining the demand for health tourism. *Tourism Recreation Research*, 40(2), 205–219. https://doi.org/10.1080/02508281.2015.1045364

Smith, M., & Wallace, M. (2020). An analysis of key issues in spa management: Viewpoints from international industry professionals. *International Journal of Spa and Wellness*, 1–16. https://doi.org/10.1080/24721735.2020.1819706

Speier, A. (2008). Czech balneotherapy: Border medicine and health tourism. *Anthropological Journal of European Cultures*, 17(2), 145–159. https://doi.org/10.3167/ajec.2008.170210

Stevens, F., Azara, I., & Michopoulou, E. (2018). Local community attitudes and perceptions towards thermalism. *International Journal of Spa and Wellness*, 1(1), 55–68. https://doi.org/10.1080/24721735.2018.1432451

Surdu, O., Tuta, L. A., Surdu, T. V., Surdu, M., & Mihailov, C. I. (2015). Sustainable development of balneotherapy/thermalisme in Romania. *Journal of Environmental Protection and Ecology*, 16(4), 1440–1446.

Sziva, I., Balázs, O., Michalkó, G., Kiss, K., Puczkó, L., Smith, M., Apró, E. (2017). Balkan region focusing on health tourism. *GeoJournal of Tourism and Geosites*, 19(1), 61–69.

Szromek, A. R. (2020). Model of business relations in spa tourism enterprises and their business environment. *Sustainability*, 12(12). https://doi.org/10.3390/SU12124941

Szromek, A. R. (2021). The sustainable business model of spa tourism enterprise – results of research carried out in Poland. *Journal of Open Innovation: Technology, Market, and Complexity*, 7(1), 1–20. https://doi.org/10.3390/joitmc7010073

Toro, F. J. & Pinos, A. 2022. *Post-tourism. Encyclopedia of Tourism Management and Marketing*. Edward Elgar (online). https://doi.org/10.4337/9781800377486.post.tourism

Valeriani, F., Margarucci, L. M., & Romano Spica, V. (2018). Recreational use of spa thermal waters: Criticisms and perspectives for innovative treatments. *International Journal of Environmental Research and Public Health*, 15(12), 2675. https://doi.org/10.3390/ijerph15122675

Van Tubergen, A., & van der Linden, S. (2002). A brief history of spa therapy. *Annals of the Rheumatic Diseases*, 61(3), 273. https://doi.org/10.1136/ard.61.3.273

Index

Pages in *italics* refer to figures and pages in **bold** refer to tables.